D1396871

THE
A PRIMER FOR INDIVIDUALS
DRONE
AND THE ENTERPRISE
AGE

DAVID
PREZNUK

PUBLISHING

The Drone Age – A Primer for Individuals and the Enterprise
© 2016 by David Preznuk

Cover design by Brian Zuckerman

Editing, book design and production by John Everett Button

Assistant editing and book design by Bethany Martinez

Manuscript formatting and graphics by Julius Broqueza

Published by Milton Chadwick & Waters Publishing
5008 Shortgrass lane, Haymarket, VA, 20169
www.MiltonChadwickandWaters.com

ATTENTION CORPORATIONS, UNIVERSITIES, COLLEGES, & PROFESSIONAL ORGANIZATIONS: Quantity discounts are available on bulk purchases of this book for educational purposes, gifts, or as a premium for increasing magazine subscriptions or renewals. Special books or book excerpts can also be created to fit specific needs. For information, please contact contact@MiltonChadwickAndWaters.com.

First published by
Milton Chadwick & Waters Publishing
February 2016

ISBN: 978-0-9884542-5-5 (hc)
ISBN: 978-0-9884542-6-2 (sc)
ISBN: 978-0-9884542-7-9 (e)

Library of Congress Control Number: 2016900796

PUBLISHING

Milton Chadwick & Waters Publishing
www.MiltonChadwickandWaters.com

This book is dedicated to
all those who endeavor
to enter and pursue
the Drone Age.

Acknowledgments

I have come a long way in my life and I would be remiss if I didn't acknowledge the special people in my life that helped make this book a reality. First I would like to thank my parents, Paul & Eva for instilling in me a moral and ethical compass that allows me to persevere when fear, uncertainty and doubt are ever present. This book would never have gotten in front of you without the passion and unwavering belief in me as an author that has been bestowed upon me by the team at Milton Chadwick and Waters Publishing, led by the man himself, John Everett Button. Without his years of encouragement, this never would have happened. Next, I have to thank the love of my life, Bonnie. She has been a pillar of support and love through this entire process. She has helped me through the many sleepless nights, contributed to the many revisions and continues to believe in me. Finally, none of this would be worthwhile without the constant love and inspiration from my son, Jack. He is a beacon of joy and a constant reminder of the wonder and awe of life. Thank you all!

About the Author

An international drone & UAS industry expert, author, and speaker, David Preznuk is the founder of Aerial Strategies, which specializes in helping organizations safely adopt, integrate, and use drones. David Preznuk is truly a visionary in the UAS industry and has been at the forefront of the migration of these platforms to the commercial and public sectors. He is extremely knowledgeable about the range of platform types and their capabilities and limitations, as well as system-related considerations – all based on first-hand operational experience. He is also well versed in the regulatory environment and market opportunities, including a broad spectrum of applications he regularly advises clients on. He has held a private pilot license for more than 25 years and has been formally recognized by the FAA for his expertise. David currently lives in the Washington D.C. area and can be contacted on his website at www.davidpreznuk.com

Table of Contents

PREFACE

I have been fascinated with the skies and space for as long as I can remember. I grew up during the great space race and remember sitting in front of the television-watching rocket launches and moon missions. When the space shuttle came to be, I remember racing home from school to watch live broadcasts of missions. I also remember with pride the final mission when the space shuttle flew around Washington DC atop a 747 on its way to the Smithsonian, where it is now on display for the world to enjoy.

I always wanted to fly, and I obtained my private pilot's license when I was in my early twenties. Even today when I go flying I have the same feeling of freedom and joy when at the controls. The sense of freedom and confidence that comes when making a good landing is something I will always appreciate.

My professional career was in technology, information technology (IT), and my personal interests were always outdoors, in nature. Whether it was hiking, biking, skiing, mountain climbing, I just wanted to be outside doing something. One day I was reading about commercial drones, and as I read I became enthralled and excited about the convergence of technologies and how they were being fashioned together to create small unmanned aerial vehicles (UAVs). I had played with R/C planes in the past, but always crashed them for one reason or another. These new UAVs were different, very different. The more I read the more I knew I had to get into this industry. I abandoned my professional career and dove headfirst into the deep end of this new industry and created Aerial Strategies LLC – a commercial drone company focused on providing services to commercial and industrial customers. The first generation UAVs (I prefer to call them platforms since they can carry an assortment of sensors and payloads) were amazing at the time, but when compared to present day UAVs I realized how fast things are changing in this industry.

The industry then and even today is full of hype. Finding useful information was difficult. *The Drone Age* is the book I wish was available when I got into the industry. Whether you are looking to get into the industry yourself, have deep curiosity about the industry, or are grappling with integrating UAV's or their content into your enterprise, this book will provide a wealth of information to help guide you through what they are, what they can do, and what you need to know about the regulations regarding their use. This is truly a primer that will give you a great foundation to inform yourself when you endeavor to use them.

INTRODUCTION

Unmanned aerial systems (UAS) or drones come in many shapes and sizes. They are used by small children for fun, and by industries for commercial purposes, and by countries to defend and protect national interests. *The Drone Age – A Primer for Individuals and the Enterprise* is intended to inform and provide a broad and comprehensive look at how the commercial UAS industry is evolving and how it is poised to impact our daily lives. Globally, the commercial drone industry is growing rapidly, and governments are under great pressure to establish policies and regulations that allow a commercial UAS to occupy airspace that is already congested.

The Drone Age will take the reader through the history of drones and touch on how they evolved over time. We will then look at the five types of drones to show some of their many uses and applications in the commercial and industrial sectors. Before using any UAS, it is important to understand the regulatory environment; we will look at the current regulations in the U.S. and discuss how the government classifies their use. Then we will take a deeper look at how these amazing little machines are being used today for commercial purposes and how they are benefiting several industries. Using a UAS seems easy in some ways, but in order to integrate them into any organization, there are many facets to consider before acquiring one for the company. We will delve into business adoption and integration issues and challenges and what to consider when implementing a drone program. Finally, we will take a look at the future and how drones could evolve over the coming years.

This book is intended to inform the reader solely about the commercial drone industry. It does not cover military or Department of Defense (DOD) drones that are used to support warfare or combat actions.

When I began writing *The Drone Age,* I started writing as much as I could about the technologies and components involved in commercial drones. I soon realized that this took me far too deep into technical jargon that might be obsolete by the time the book is published. It also meant the book would be 300-500 pages. Instead, I chose to elevate the conversation and avoid comparing technical components (Pixhawk vs NAZA) and trying to convince the reader of the merits of one system over the other. This book is ideally suited for someone that wants to go beyond the headlines and media reports about drones and be better informed on the subject. Also, if an organization is seeking to introduce drones into their organization, this should be considered a must-read first book. We will touch on many of the challenges and issues any organization will face when looking to adopt these technologies.

Thank you for picking up *The Drone Age*. I hope you enjoy reading it as much as I did writing it.

THE
BASICS

PART 1

A BRIEF HISTORY OF UAS EVOLUTION

What is a Drone?

The Drone Age is upon us! As unmanned aerial vehicles or "drones" enter and proliferate into and throughout our everyday lives, many questions and some fears arise. There is no doubt that drones are arriving and will be here in one shape or another from now on. Looking at the history of how any technology evolves, we know that as a technology is used, it improves, evolves, and infiltrates other facets of our lives and businesses. This is an incremental and natural evolutionary step in nearly all product development lifecycles. Once the product, in this case, a "drone" is used, it becomes better understood, reliability and quality improve, and it will become a part of our everyday lives in new and imaginative ways.

Since the Industrial Revolution (I and II) began we have experienced an increase in the velocity of change associated with the advancement of new technologies. We witnessed this with the advent of the cotton gin that we learned about in elementary school, the locomotive, the automobile, telecommunications, and most recently the Internet. Of course, there are many, many other examples we can point to. In our case, we will explore the possibilities, the pace (velocity) in which change and innovation is occurring, and how fast it translates to real benefits to society, industry, and humanity.

Looking back through my own life I have experienced many advances that I was initially pessimistic about then quickly couldn't live without. I was born in the 60's and have had the benefit and luxury of all the things that have changed in the relatively short time I have been alive. An easy example is the telephone. When I was growing up it was reserved for adults and their important conversations. Calls were brief so as not to tie up the line. There was no call waiting, answering machines, caller ID, or any other capability yet, but they all had their heyday in the 1990's and early 2000's. Today homes are increasingly moving away from dedicated "land-lines" in favor of the omnipresent and immediate access afforded by a cellphone –an amazing little computer with caller ID, multiple lines, call forwarding/blocking/special ringtones, etc. etc. etc. Today calls are being replaced with text messages and tweets.

I believe drones will evolve similarly to the cellphone. I can envision a day in the not-too-distant future when we as a global society are reliant on drones for many tasks we don't even know we need yet! Unmanned systems have been around for a long time, they have been used in our oceans, on land, and in space. Our airspace is a natural next step. You may not share this view now, but I hope by the end of this book you may have a new perspective. Before we dive in, let me jump ahead a little and share a tantalizing scenario that may be possible in the near future. I can envision broad, spectacular and amazing applications that will better our planet, and by extension, our lives using drones.

We all know our atmosphere is polluted. Imagine a day when unmanned aerial platforms will be autonomously flying in the airspace around the world with the sole mission of reducing ozone or CO_2 to optimum levels for vegetation in all regions of the world. Yes, I believe this is possible, but how and when? I believe this will be possible with the convergence of artificial intelligence, quantum computing, and nanotechnology. All of these are advancing at blistering speeds. One day, these will be combined in a fashion that allows them to work collaboratively to achieve a common goal. A group of drones – swarm, fleet, or flock – whatever we want to call it, would have the computing, analysis, and chemical compounds (in this case nanotechnology) to monitor, analyze, and fix certain conditions. In this example, the swarm scrubs the atmosphere and removes excess ozone and CO_2.

Furthermore, each drone in our imaginary fleet could fly for six months using solar and battery technology; the fleet is configured in such a way that it will fly indefinitely and new drones will replace obsolete, retired, or spent (used up) drones autonomously. Each drone is equipped with electronic sensors that allow it to broadcast its location, coordinate and avoid other aircraft (no possible accidents), and to conduct its mission. As the fleet flies, it is constantly analyzing the levels of ozone and CO_2 and dispersing a real-time molecular fix, reducing the levels of ozone and CO_2 in the atmosphere. This scenario is no longer sci-fi, it is within reality's grip. Before we delve too far into capabilities and the future, let's begin with understanding the basics.

Drone / UAS / UAV / RPV – all refer to Unmanned Aerial Systems

Drone, unmanned aerial vehicle (UAV), unmanned aerial system (UAS), remotely piloted vehicle (RPV) all refer to the same thing: an aircraft that is flown and/or controlled without an onboard pilot. For the purposes of this book, we will use "Drone", "UAS", and 'UAV" interchangeably as the term to describe all of these aerial vehicles. There are some slight differences we will explore later in this book.

These terms are also evolving and expanding as the commercial industry increases its understanding and use of UASs. For example, autonomous flight is now something that is becoming a common term used in the industry, but its use(s) may and often refer to different aspects. Some use autonomous to refer to the UAS flying entirely on its own without intervention from a remote pilot. Presently, even a low cost ($1000) prosumer grade UAS may be pre-programmed to fly a pre-determined flight plan, all the pilot needs to do is hit the launch button and the UAS will fly an entire pre-programmed mission then land without any input or instruction from the ground-based pilot. The remote pilot or operator may intervene at any time to correct the flight plan, re-program some aspect of the flight, abort the flight or take over control and fly the remainder of the mission from the ground. Even though many call this autonomous flight, they are actually referring to a flight of pre-programmed specific instructions.

True autonomous flight is something that is technically feasible and achievable, but lacks the legal and regulatory guidelines to be implemented. True autonomous flight would mean that the UAS, once programmed with a task or mission, would complete that task or mission entirely on its own without input from the ground. For example, if the UAS determined, on its own, that the winds were more favorable at another altitude, or it calculated a more efficient set of waypoints that avoided potential obstructions or something that could introduce risk into the mission, it would determine the best way to complete the task or mission. Of course there are benefits and risks to autonomous flights, benefits may include higher efficiency, less potential for human error to be introduced. However, once the UAS has the ability to complete a task or mission on it own, it may make changes that introduce problems or risks elsewhere (general aviation, commercial aviation, industrial complexes, special events, and the list goes on-and-on).

Until recently, anything unmanned was thought to be a military or DOD drone flying over a hostile theatre. In these cases the drone would likely be weaponized, very large and an instrument of destruction. With new capabilities and the rapid advancements in technologies this has all changed. No one thought a hobbyist's remote controlled (RC) airplane or helicopter could be considered a drone. And public opinion and regulations have not yet adjusted to the changes. The lines have blurred between what is recreational use and what is commercial use. This is an issue that raises many questions we will discuss later in the regulatory section.

How did they come to be?

The term "drone" was borrowed from the natural world. It refers to the male honeybee and he has no stinger (or armament). The drone's only purpose in nature is to mate with a queen bee, and once this has been done the drone no longer serves a purpose and dies.

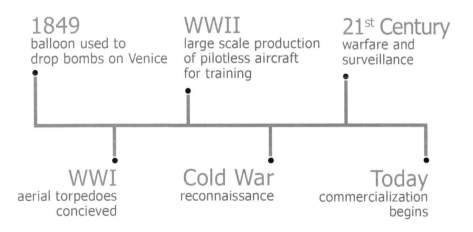

Brief History of UAS

Figure 1: History of Drones / UAS

The earliest recorded use of an unmanned aerial vehicle for warfare occurred on August 22, 1849, when the Austrians attacked the Italian city of Venice with unmanned balloons loaded with explosives. The concept of pilotless aircraft was introduced during World War I (WWI) in the form of aerial torpedoes, which were primitive versions of today's cruise missiles. Although the concept was successfully demonstrated to the U.S. military, the product could not be fully developed and produced before the war ended.

During World War II (WWII) pilotless aircraft were mass produced and successfully used by the military for target practice. Ground-based anti-aircraft gunners honed their skills using remotely controlled aircraft. During training WWII airplane pilots also used pilotless aircraft for air-to-air target practice, increasing their skills during real air-to-air dogfights against the enemy. The post-WWII era saw advances in technology and capabilities that moved pilotless aircraft from the training arena to the battlefield. As technologies advanced and improved, pilotless aircraft were transformed into reconnaissance platforms to gather intelligence. They were used during the Cold War period and during the conflicts in Korea and Vietnam. During the early days of drone use, they were often considered unreliable and expensive.

Research and development (R&D) by government agencies (the military, NASA, and intelligence agencies), and commercial industries expanded during the 1970's. By the 1980's drones were used by the military for real-time reconnaissance, electronic jamming, and they were beginning to be armed with precision munitions. NASA, NOAA, and other scientific agencies were also using these technologies and capabilities for scientific purposes.

Today UAS technologies are being used for military, scientific, commercial, and industrial applications all over the world. Technological advances have permitted highly sophisticated electronics and computing power to be squeezed into smaller and smaller components. These advances allow even the smallest unmanned aerial vehicle to carry satellite navigation, accelerometers, electronic compasses, on-board stability control software, and a whole host of other specialized monitoring sensors to permit the safe operation of the UAS. These are not the remote-controlled (R/C) toys some of us think about when we see small remotely piloted aerial vehicles. These are small technological marvels that have impressive capabilities, benefits, and applications.

Commercialization begins

In February of 2012 the U.S. Congress approved the reauthorization bill for the Federal Aviation Administration [FAA]. This bill had a provision that required the FAA to "open-up" the US airspace to commercial unmanned aerial vehicles by September of 2015. The bill also required that the FAA create six test sites across the nation to determine the best way to integrate drones into the National Airspace System [NAS]. With this congressional action, the US commercial drone race began.

Commercial drone use has been a hot topic ever since. Safety, security, and privacy concerns have been a mainstay in the media and headlines in the years since the congressional announcement. Sadly, initial public and media sentiment was pessimistic and negative. Most of this I attribute to the conventional wisdom at the time and to human nature's resistance to change.

The September 30, 2015 deadline has come and gone and the FAA has made significant progress towards realizing the congressional mandate of February 2012, but it still has a ways to go. We will discuss this in more detail in Chapter 5 (US Regulatory Environment).

TYPES OF DRONE PLATFORMS

Figure 2: Military Drone – Predator

Drones come in all shapes, sizes and configurations. The idea of drones conjures up various images in our minds, from radical military weapons to small toys children fly in the living room. By now I am sure many people have seen military drones like the Predator, which is used in direct combat theatres or, more recently, they've seen a small multi-rotor like the DJI Phantom 3 that has been in the news. The predator is a full-sized aircraft and would be more aptly compared to a modern Air Force fighter plane. It is large, requires a lot of specialized equipment and personnel to operate. Multi-rotor drones like the DJI Phantom 3 are extremely complex electrical drones requiring specialized firmware and software in order to maintain stable flight.

In addition to these two types of drones, there are other design platforms in existence, each with their own advantages and drawbacks. It's important for the consumer/organization to understand the dif-

Figure 3: DJI Phantom 3

ferences between the types of drones, and how they are able to accomplish an individual's needs.

The DJI Phantom 3 is a multi-rotor drone. Multi-rotor drones have several individually controlled propellers attached to their own motor. Each motor is controlled independently from the other motors on the drone. The collection of motors (as few as 3 or as many as 12 or more) each generates the required thrust to lift the drone and vary propeller speed to control the stability of flight and the direction of flight.

UAS Components

All drones require a few common components to function: an airframe, some sort of propulsion [motor], a control system and a communications system. Also, to use a drone to perform any task other than flying, a sensor [payload] of some sort is attached to the airframe. The most common payload is a camera capable of shooting video and still pictures. Additional sensor and imaging options may include thermal, Infrared [IR], multi-spectrum, hyper-spectrum, and LiDar (Light Detection and Ranging), to name a few. Finally, once the UAS has completed its mission, some form of post-processing needs to be performed to translate the data captured during flight into consumable information for the intended audience.

We'll devote this chapter exclusively to a discussion of the types of UAS platform designs available today: fixed-wing, rotary-wing, multi-rotor, lighter-than-air, and tethered.

Components of a UAS

COMPONENT	TECHNOLOGY
AIRFRAME	Platform designs - fixed wing, rotary wing, multi-rotor, lighter-than-air, tethered
POWER/PROPULSION	Electric, gas, turbine, hybrid
SENSOR/PAYLOAD	Camera, thermal, infrared (IR), LiDar, other
COMMUNICATIONS	RF transmitter, ground station, hybrid
CONTROL SYSTEMS	Hardware and software
POST-PROCESSING	Hardware and software (image, video and sensory data processing

Figure 4: Components

Airframes

Advancements in airframe designs and materials are moving at a blistering velocity. Materials that were once very expensive and exotic are finding their way into lower priced platforms all the time. Carbon fiber is becoming commonplace in many mid-level and higher end systems. Advances made upstream in DOD or commercial aviation sectors are quickly being scaled down and applied to the UAS platform.

Innovative techniques with plastics and foam airframe components are being used to increase strength and decrease weight. This approach immediately translates into improved endurance for any flying machine or system without adding more power. First generation multi-rotors from one to two years ago, for example, although similar in appearance and design are vastly improved in terms of materials and design that reduce weight. These innovations can be seen across the spectrum of platforms we discussed and the innovations seem to just keep coming.

Regardless of the type of UAS platform you may consider, several other considerations may need to be taken into account; for example, the power plant. Fixed-wing and rotary-wing platforms can accept several different power options, among them are electric, gas, turbine, or even a hybrid. Each will have its own set of benefits and liabilities. Let's say you wanted a fixed-wing platform that was small enough to travel with, had great endurance and payload capability and you decided a gas or turbine power plant was perfect for your applications. You were going to be traveling with your platform using commercial aviation to travel frequently around North America. You would quickly discover that transporting your new drone would come with a whole set of new challenges—namely, the restrictions of traveling with a volatile fuel source. Even if you didn't intend to carry any fuel with you and decided all fuel would be purchased at the intended destination. Residual fuel vapor in the engine and fuel tanks would carry restrictions of their own adding cost and operating procedure considerations that may not justify or diminish the value of initial platform selection.

For our purposes, we are going to concentrate on platforms [drones] that are under 55lbs (which is the FAA designation for small unmanned aerial systems). This category of drones is the current flashpoint of commercial UAS industry.

Types of Drone Platforms

| FIXED-WING | ROTARY-WING | MULTI-ROTOR | LIGHTER-THAN-AIR | TETHERED |

Figure 5: Types of Drone Platforms

FIXED-WING

A Fixed-wing platform is a small or large airplane, depending on its scale. Some can be as large as a fighter jet and others can be as small as a hummingbird. These types of drones have the same attributes and benefits as their much larger siblings. They are well proven and offer an abundance of configurations, multiple power plant options as well as payload / sensor possibilities. Fixed-wing platform types are the most aerodynamically efficient platform of all the five types. The power-to-weight ratio is impressive and offers the operator a broad spectrum of possibilities including interchangeable power plants, sensors, launch and recovery equipment, command and control systems, and many more. It is easy to modularize a single platform to meet the goals of many missions and tasks.

A fixed-wing platform has significant advantages for many applications when comparing endurance, range, and payload or sensor options. For example, let's say we needed photos and IR information for a large agricultural production, thousands of acres. The fixed wing UAS would be the ideal selection because of its endurance, variable payload and sensor options, and its ability to be deployed and retrieved easily in open areas.

In this agricultural example, the flight profile is straightforward, and due to the capabilities integrated into all modern platforms, the flight plan can be saved and the exact same profile can be flown repeatedly in the future. This would provide the farmer with a rich set of data that could be a baseline to determine the most effective farming strategies to increase yields and reduce over fertilizing and irrigation. These basic steps are win/win for the farmer and for the environment. The farm/ranch will reduce real costs— fuel, irrigation, fertilizer, pesticides and time —as well as reducing the negative environmental impacts associated with the overuse of fertilizer, pesticides, and excessive irrigation.

Technologies are to the point where the aerial data can immediately be downloaded into the farm equipment computer and be put to use in minutes. Once a baseline has been established the farmer could then overlay other valuable information / data whether it be historical, current or projected. This data could be soil sample information over the past number of years, historical yield, market price, weather, or a whole host of data that would further allow the farmer to manage crops and predict the health and yield of the land. Taken a step further, the farmer would be able to place the exact amount of starter fertilizer blend on each seed based on the soil composition and moisture content of the precise location of each seed. Yield losses would decrease, and the land would be better managed and preserved.

Figure 6: Launch

Of course, with the good comes the bad, so let's look at some drawbacks of this platform. Fixed-wing platforms clearly have a place in industry and will for the foreseeable future. One of the potential drawbacks of the fixed-wing platform is its launch and recov-

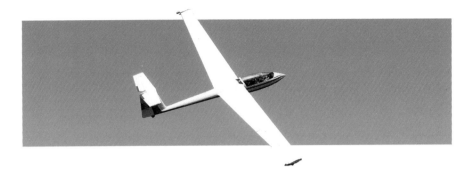

Figure 7: Fixed Wing

ery requirement. Fixed-wing platforms are not able to takeoff or land vertically and thus require some horizontal ground distance. Depending on the size of the platform, this ground distance may require a paved runway, a catapult system, or if small enough, may be deployed by hand by throwing. When ground distance or clearance is required, obstructions like trees, terrain, buildings and people must be accounted for and avoided during the launch and recovery phases of the flight.

Landing or "recovery" procedures will vary by platform. Some will execute a controlled crash into the ground and, intentionally, break into several pieces that may then be reassembled for the next flight. Some have reinforced components and are intended to skid across the ground and remain completely intact. Others require an improved surface (grass, dirt, or pavement) or some sort of containment systems, i.e. a net, or some other means of bringing the platform safely to a stop. Recovery of a fixed-wing has its own skillset more akin to piloting a plane. Regardless of method, landing a fixed-wing platform, autonomously or manually, requires planning, vigilance, and oversight to ensure good operating standards and to reduce risk.

Launch and recovery are considerations that should be weighted according to the needs of the majority of your mission / project profiles. If you will be using the UAS in rugged terrain or areas with many obstructions, be sure to have a clear understanding of the capabilities and limitations before making a selection. Also, be aware that in high altitudes takeoff and landing distances will increase for drones the same as they will for full-sized airplanes.

ROTARY-WING

Rotary-wing platforms are, for all purposes, a scaled-down helicopter with all the same mechanical complexities. These types of drones are the most complex mechanical platforms of the five types we are discussing in this book. Rotary-wing drones, like their full sized brethren, require a significant amount of power to takeoff and to maintain controlled flight. Rotary-wing airframes also generate a significant amount of vibration that must be dampened to ensure any sensor payload like video, photography, IR or other imaging data are isolated from this vibration. Otherwise, the motion would cause image distortion, rendering the sensor data unusable. However, if the payload is not extremely sensitive to vibration, for example, an agricultural crop dusting application, then vibration may be less of concern.

Additionally, the control systems of rotary-wing platforms require vigorous maintenance before and after flights to ensure no damage or integrity issues with control surfaces, components or the airframe itself occurs. For these reasons, this type of platform would fall into the category of high cost of ownership.

That said the rotary-wing has some impressive capabilities that serve many applications quite well. The rotary-wing platform has vertical takeoff and landing (VTOL) capability. Also due to airframe design, this type of platform can be fitted for many different power plants, even on the same airframe. Rotary-wings can be powered with electric, gas (two-stroke or four-stroke), turbine, and even hybrid power plants. This provides the operator a high degree of flexibility and configuration depending on the flight profile. Additionally, it provides the ability

Figure 8: Rotary Wing

to have purpose-built power plants optimized for a specific application that balances endurance and payload. Also, the airframe itself is very versatile and may be fitted with various payload modules to serve its application.

The rotary-wing platform has been used outside the continental United States for agricultural applications in rugged terrain, and they enjoy a loyal following. Imagine trying to fertilize or apply pesticide in steep, rugged, mountainous terrain found throughout Central America and Asia. Due to their flexibility and various power plant options it is possible to configure a rotary-wing platform capable of carrying heavy payloads of hundreds of pounds for extended periods of time (1-3+ hours).

Another advantage is that rotary wings are able to fly in all types of weather conditions, including moderate or even high winds, while maintaining a stable platform for its intended purpose. This is an important consideration when flying in regions where weather, and especially wind, can vary wildly.

MULTI-ROTOR

The multi-rotor platform is likely the one that most people are familiar with. This type of platform has drawn the most media attention over

Figure 9: Multi Rotor

the past one to two years. They are readily available, come in all shapes and sizes, and have extremely impressive capabilities all for very attainable prices. A ready-to-fly (RTF) multi-rotor with a high-definition camera and video camera can be purchased for less than $1000. This platform is packed full of very sophisticated electronics that enable the most reluctant operator to feel comfortable flying within minutes.

The ease of use is one of the reasons this industry continues to expand at unprecedented rates. An off-the-shelf ready-to-fly multi-rotor from 3D Robotics, DJI or any number of vendors, can have you home and flying in no time. These, like all types, have some of the most advanced electronics available in the market, which allow the platform to maintain stable, and controlled flight by using powerful processors taking in telemetry, GPS, and control input data. It processes all the data on board the platform 100's or 1000's of times per second. This frees the operator to focus solely on navigating the platform to a precise location without having to compensate for wind or other forces.

When the operator moves the control sticks to initiate a move, the multi-rotor will automatically adjust the propeller speed of each motor to move in the direction indicated. During training sessions, in a strictly controlled environment, I demonstrate what happens when you turn off all these electronics and attempt to fly one of these in full manual mode. It is very difficult, even for an experienced operator. I do not encourage this exercise for it immediately demonstrates how much workload the platforms remove from the operator!

A multi-rotor can have many individual motors; the most common is a 4-way configuration, but it is not unusual to see 6, 8, 10 or more on an airframe. More motors require more power and a heavier airframe to support all the dynamics at work during flight. Multi-rotors are the most agile of all type of platforms and are extremely stable in all possible configurations. This type is very versatile and great for inspections that require close proximity. The multi-rotor is also multifaceted and can be adapted to carry several payload configurations. Currently, multi-rotors are solely powered by electric motors, which does limit their endurance and payload capacity. This type is the most complex electronic type of platform since all control systems are electronically manipulated and require firmware and software to maintain stable and controlled flight.

Due to the wide availability and low entry price, this type of platform has become the most popular type of platform around the globe. It is estimated that the US market will add nearly one million new multi-rotor platforms to it economy during the 2015 holiday season.

LIGHTER-THAN-AIR

The lighter-than-air (blimp) platform type is a lesser-known platform but has been around for a long time and has many commercial, industrial and Department of Defense (DOD) applications. Smaller scale blimps have become increasingly popular in large indoor venues. These platforms are very stable and have a robust payload capacity and good endurance, depending on the type of gas being used to float the blimp. For applications that require long endurance and docile

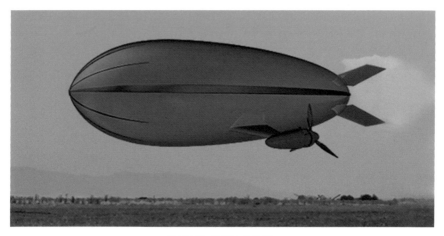

Figure 10: Lighter than Air

maneuverability this platform really shines. Although the platform is simple in construction and control systems, the complexity increases with the type of gas being used to maintain lift. If using specialty or exotic gases to generate lift, the operator may be required to obtain a special use permit, and enact specialized handling and operating procedures including a support team. These types are most susceptible to weather issues such as high or variable winds due to their volume and limited controllability.

TETHERED

The last type of UAS we are going to discuss is the tethered platform. The tethered platform, like the lighter-than-air, is a lesser-known and lesser-used platform, but like the blimp it has several functions in industrial, commercial and DOD applications. The tethered platform is typically a blimp or a multi-rotor platform that is connected to the ground using a tether. The most beneficial aspect of a tethered platform is that you can use the tether to send power to the motors and to connect to the sensors that are being carried by the airborne platform. In the case of a multi-rotor configuration, the endurance is extended to the limits of the power being supplied from below. The mobility of tethered platforms is limited to the ability to move the tether and support system. A platform of this type can move freely within the range of the tether, provided the tether is unobstructed.

An industrial, commercial, or DOD application that requires loitering for long periods of time but can also be easily relocated is an ideal application for a tether platform. A tether platform can work nicely for security purposes, for example. Let's consider the following scenario: a busy international port on the east coast of the US needing to augment its existing security and surveillance capability due to a heightened level of security. In this scenario the threat is unknown and the duration of the additional monitoring is anticipated to be 30-60 days. It would be difficult, time-consuming, and expensive to integrate new static (fixed position) security capability into their system. Also, the nature of the threat is very broad and requires a need to constantly change what and how they are monitoring a large area. Adding a combination of lighter-than-air and multi-rotors with payloads that provide a broad range of monitoring capability – chemical monitors, IR, thermal, high-definition video and a host of other capabilities – may need to be considered.

Figure 11: Tethered

In this case, it would be easy to redeploy a tethered system to other areas in the threat zone or leave them in place for an extended period of time. Also, should the threat change or become more clear, additional or specialized monitoring sensors could easily be added to the payload of the tethered system. This is a very flexible system and can be deployed in a wide range of weather conditions while maintaining operational readiness.

When considering the many types of UASs, it helps to realize all systems have distinct capabilities and limitations and some are broader than others. Regardless of the type of platform you select, the important thing to factor into any decision is the fit for purpose. No single platform will likely meet 100% of your needs, but with some careful consideration, a UAS with specific payload/sensor options can justify the cost of operating the system, which leads us nicely into the next section.

Comparative Analysis Framework (CAF)

Before embarking down the drone path, whatever your UAS goals are, for optimal results it is vital to have a clear understanding of what purpose or need the drone is fulfilling. A quick decision, what I refer to as "buying on headlines", often ends badly. Let's look at a recent example involving the FBI, who were too quick in their purchasing of a UAS fleet. While they had clear, perceived expectations from the capabilities provided by a UAS, their uninformed actions were ultimately wasteful.

Don't be the FBI or ATF![1] These two agencies saw what they perceived as clear benefits of using drones to support their respective missions. With good intentions, the FBI and the ATF spent $2.3 million on 23

---

1. See this article "FBI and ATF spent $2.1 million on 23 drones that don't work," http://arstechnica.com/tech-policy/2015/03/fbi-and-atf-spent-2-1-million-on-23-drones-that-dont-work/

drones that they cannot use! The UASs sat on the shelf for many rea-sons – they didn't have knowledgeable operators [pilots] to fly them, they didn't have FAA authorization (COA) to fly them, the selected UASs didn't have the endurance required, and the list goes on-and-on. This is the classic example of not having clear expectations or require-ments for the intended use. In order for any governmental agency or commercial enterprise to operate drones, they must meet FAA regu-latory compliance requirements, have qualified operators (pilots and payload/sensor operators), be able to maintain the fleet, and have standard operating procedures (SOPs).

The moral of this story is this: Don't buy on headlines and do your homework before deciding which platform is best for your needs. In many cases, you will not be able to find a single drone that meets 100% of your needs, which is ok. Having one that meets a fraction of your business needs is sufficient provided you have the analysis to support that need.

Let's take this a step further; perhaps you have a need for infrared [IR] data for agricultural survey purposes. It would be easy to jump to the conclusion that a fixed-wing platform would be best due to its endur-ance, range, payload, low complexity and cost of ownership. Howev-er, if the agricultural land was in steep terrain or was geographically separated, then a fixed-wing could find itself operating the majority of its time around its limitations or weaknesses instead of its strengths. Given the terrain requirements, in this case, a fixed-wing would strug-gle. Perhaps a rotary-wing or multi-rotor would be better suited to the needs in this instance. Another option might be to use multiple platforms. We could then consider a fleet that is comprised of any or all types to meet the demands of our application(s).

Having a clear idea up front of what and how a drone would be used is vital to its success. In the FBI / ATF example, they quickly realized they had wasted their money and ended up giving the drones away. The outcome was lose-lose, they didn't have a platform that met their needs and are now reluctant to include drones in their agencies. Had they allocated even a small fraction of their budget to determining their requirements and researching models then the results may have been better.

Comparative Analysis Framework

CAF	FIXED-WING	ROTARY-WING	MULTI-ROTOR	LIGHTER-THAN-AIR	TETHERED
ENDURANCE	High/Excellent	Variable	Low/Moderate	Excellent	Excellent/Extreme
PAYLOAD	Moderate/High	Moderate/High	Moderate	Moderate	High
POWER SYSTEM	High	High	Low	Low	Moderate
LAUNCH & RECOVERY	Low	High/VTOL	High/VTOL	Moderate	Low
PLUG & PLAY	Moderate	Moderate	Moderate	Moderate	High
MAINTENANCE	Low (less)	High (more)	Moderate	Moderate	Moderate
COST OF OWNERSHIP	Moderate	High	Moderate	High	High
COMPLEXITY	Moderate	High/Extreme	High	Moderate	High
AGILITY	Moderate	High/Extreme	High/Extreme	Moderate	Low
LOITER-ABILITY	Low	High/Extreme	High/Extreme	High	High/Extreme

Figure 12: Comparative Analysis Framework

Figure 12 shows fit for purpose. I have done several assessments and used this framework to quickly consider what platforms may be worth considering for a given application. The criteria may be adjusted to meet a particular need; it may not always be about endurance, range, and payload capacity. For example, if my need is to take pictures of residential real estate to create marketing brochures, then endurance may not be a factor at all, since my flights may be limited to 5-10 minutes. Even if they needed to be longer, I could easily land to change batteries and resume taking pictures or video. In this example, I could fulfill 80-90% of my needs with a small investment and start putting it to use immediately – provided I was regulatory compliant. Once I had this platform in use, I could then determine if other more robust platforms were necessary; or I may expand my service offering as my expertise evolved.

OTHER COMPONENTS OF A DRONE

Now that we've covered the types of drone platforms available to choose from, we'll explore the other components of a drone/UAS that are needed for it to function. This topic could be a series of highly technical books all by itself. We are not going to dig deep into the bits and bytes of all the components, but rather, we will discuss them at a sufficient level to have a common understanding of what they entail and how they interconnect and operate to provide the capabilities necessary for a drone/UAS to do what it does.

Propulsion / Power

As with airframes, fuel-based systems (2-stroke, 4-stroke, or turbine) are very mature technologies and are continually being tweaked to improve performance. Power plants for a UAS have undergone massive changes recently. The big leaps are occurring in the electric-based power systems.

Let's look at the electric motors used for any of the platforms we discussed; the combination of motor design and power management has seen amazing advances in just the past year. The new motors are impressively designed to make the most of new advances in battery technology. For comparison, my first generation multi-rotor purchased in early 2014 had a flight time of approximately eight minutes under ideal weather conditions. This first generation platform had motors that were inefficient and operated at or near 100% of their maximum power for normal operations, leaving zero power in reserve extenuating or emergency situations. Additionally, the motors would wear out fast because they were operating at their maximum designed operating envelope. Later generation motors are now designed to operate at less than 100% of their maximum power rating. Which translates to a significant reduction in power consumption and an increase in platform endurance.

Sensor / Payload

Sensor and payload are two different terms. The definition of a sensor is a device that detects or measures a physical property and records, indicates, or otherwise responds to it. Examples of sensors are infrared, thermal, multi/hyper-spectrum, or a camera for taking video or photographs. The payload is the carrying capacity of the aircraft, usually represented in terms of weight. For the vast majority of the current consumer UAS platforms, the sensors are often a point-of-view (POV) camera attached to the UAS for the purposes of taking pictures or videos. As the size of the UAS platform increases so does the payload. A larger platform has more payload capacity and can carry more fuel or sensors. It is becoming more common for a drone to carry more than one sensor at a time – a high definition camera for capturing video or still photos and an IR or thermal camera for capturing non-visual data. I believe it will soon become possible to purchase a platform that is modular and can easily swap sensors for different mission-specific applications.

The final component to mention is the gimbal. A gimbal is a device that allows a sensor to remain in consistent orientation independent from the vessel carrying it. On UASs the gimbal isolates the platform vibration from the sensor it is carrying, thereby reducing transmission of vibration to the sensor which may render any captured data useless due to blur or vibration static.

VIDEO

This is an area that is rapidly changing. Video quality on a commercial UAS has improved greatly in the just the past year and a half. Video shot in 4K is now becoming the defacto standard for video content. So the money spent on camera technologies in other industries has found its way to the drone and is contributing to the development of next generation digital sensor capabilities. Point-of-view cameras were the standard in early UAS generations because they were small, light and took decent images. Now purpose-built cameras and even large commercial video cameras can be fitted onto a UAS for shooting amazing videos. A feature film digital camera (Red One) for example is now routinely flown on a UAS platform to capture stunning aerial shots for movies. Helicopters and planes are no longer necessary. You may have noticed aerial shots in recent movies, commercials and TV shows; the majority of those shots were most likely captured using a drone.

PHOTOGRAPHY

Like video-capturing technologies, photography and how we use pho-tography, using a UAS is changing. First generation aerial cameras did not produce the high-quality images we get today. Early cameras were usually point-of-view cameras having the drawback of a large field of view (FOV), which created distortion at the edges of the images and also had a curvature of the earth effect. The curvature is not as appar-ent from the ground but becomes apparent with aerial photography.

Aerial pictures from the most current generation cameras are vastly improved in terms of image quality and range of control from the ground. It is now standard to be able to control exposure, shutter speed, image sensitivity, and other options from the ground. This was a game changer. When I worked with a first generation UAS with a POV camera, I had to set all parameters from the ground, start the camera (whether it was video or time lapse for photos) and takeoff. When I landed five or 10 minutes later, I had to review the entire video to find the few clips I needed. Sometimes I had to fly again because the shot wasn't right. If I were taking photos I could have hundreds or even thousands of images to review to find "the one" or two that were perfect. It was a very time-consuming post-processing activity that I learned a lot from. Today, I can takeoff, frame the shots I want and control the entire process from the ground. I can even stream the video in real-time over the Internet for customers who need immediate access to aerial content.

INFRARED

Infrared (IR) is invisible radiant energy, that is, electromagnetic radiation with longer wavelengths than those of visible light. This spectrum is used in many specialized applications and is becoming increasingly common in agriculture and various other aerial applications due to the information garnered from this spectrum of light. In agriculture, IR applications may be used to monitor the health of a crop. If an area is showing signs of deterioration, a farmer may quickly identify problem areas, and then treat those affected areas before the problem results in widespread losses. In the past, this capability was only available by using expensive satellite or other airborne imagery. The cost and latency associated with satellite imagery limits its information to those with large budgets. Today the cost of acquiring real-time IR data is so low that all farmers are able to take advantage of this capability to better manage their crops.

THERMAL

Thermal sensors detect temperature. This basic capability has many applications in the real world and is already used in many devices we use everyday – from our cars to our computers. They have been used from the air to aid in search and rescue operations on land and oceans. Using this capability on a UAS allows many industrial and commercial applications to put this sensor to great use more easily. Some examples would be inspecting power transmission lines, solar panels, crops, and any application where temperature is vital to determining stress, fatigue or potential failures.

MULTI / HYPER-SPECTRUM

This type of sensor provides the ability to see various spectrums of light, including those invisible to the naked human eye. For example, people see the visible spectrum of light (red, green, &blue – RGB), a goldfish sees the world through infrared (IR), and a bumblebee can see in the ultraviolet spectrum. Using this sensor we can augment our sight to use these other spectrums as well. This enhancement allows us to determine mineral content, vegetation density and composition, and many other exciting differences that were previously unnoticed. This type of sensor has broad application in agriculture, oil and gas exploration, ecology, geology, oceanography, and others. Combining this sensor data with other sensor data further enriches the content and imagery allowing more precise assessment and analysis.

LIDAR

LiDar, which stands for Light Detection and Ranging, is a sensing technology that uses a laser to measure precise distances between the laser and the target. It is commonly used for making high-resolution maps often used in geology, mining, forestry, seismology, archeology, atmospheric science, surveying, agriculture and even law enforcement (to determine the speed of a vehicle). Until recently this technology and the data it provides has been expensive to acquire. Like many technological advances, mass adoption, and miniaturization has brought this capability to new industries, markets, applications, and platforms. In the past, aerial LiDar data were reserved for satellites or specialized planes or helicopters. Today a LiDar unit small enough to mount on a small commercial UAS is increasingly being used.

OTHER

Other types of sensor and sensing technologies include chemical, radiation, pressure, and more highly specialized technologies. Sensor technology is a specialized field that's finding new and exciting ways to merge UAS platforms for data acquisition. Chemical sensors are a natural fit for small UAS platforms since they are extremely portable, easily deployable, and can be managed by a single person. Having an electronic sensor that can detect multiple chemical compounds yet still fit on an off-the-shelf UAS platform is an application that will take the commercial UAS industry in new directions. New types of inspections and monitoring will emerge, and human safety will be improved in high hazard environments.

Communications

RADIO FREQUENCY (RF)

All commercial off-the-shelf UAS platforms use radio frequencies to transmit instructions to the UAS and receive telemetry and video data from the platform. The Federal Communications Commission (FCC) in the U.S. regulates radio spectrum, and the RF spectrum used for commercial UASs is limited and in demand for several services. Only a few frequencies are available to off-the-shelf UAS platforms, and all retail UAS manufacturers must use the same limited RF due to FCC restrictions. Other frequencies are available but require special licensing, and their use is tightly regulated and monitored. The RF spectrum currently used has limits for commercial applications due to the regulated power (for transmitting) restrictions imposed by the FCC. Primary limitations are range and signal bandwidth, both of which have strict compliance standards.

In densely populated areas, RF is congested and is susceptible to interference, reducing the effectiveness of RF transmissions and reducing the operating range of the UAS. The telemetry and video data may also be interrupted or severed. These RF transmissions are not secured in any way, making them vulnerable to deliberate interference (jamming/spoofing) or interception. Low power transmissions, like those with the UAS platforms we are discussing in this book, will travel great distances when unobstructed. This makes them easy targets for interception. I have spoken to customers who see this as a potential risk to operations since they potentially could lose control over this aspect of flight. Discussions of encrypting these RF signals to provide some level of security have been coming up more often.

Control Systems

All UAS platforms require hardware, software, and firmware to operate. The sophistication of these components is evolving at a rapid rate. Software and firmware are essential to the operation and control of modern UASs. These integrated control components are what makes them so capable, powerful, and easy to use. Software applications that run on smartphones, tablets, laptops, and virtually any other mobile device or desktop are now the norm. Software and firmware are responsible for the stable, controlled flight that permits some platforms to fly. Take a multi-rotor, for example. It would be impossible to safely operate one without software or firmware. The ability to have control over each motor and to maintain control for coordinated flight outdoors in 3 axes is beyond our human processing power.

Current generation software and firmware are light-years ahead of where they were only a couple of years ago. First-generation UAS platforms required rigorous vigilance, tweaking, configuring and up-grading to function. The user interface was primitive at best and there were many days lost to dealing with technical support. Today the user experience is pleasurable and makes these platforms so popular and easy to fly. The workload is now managed for you by the software or firmware. All you have to do is steer and if you don't want to steer, then you can pre-program a flight using an application, upload the flight plan to the drone and hit the launch button, then watch real-time video on your phone, tablet, or laptop. Once the UAS completes the flight plan it will land as instructed and you can start over again.

If there is a drawback to all this it is keeping the software and firmware current and up-to-date. Failing to do so may result in operational issues or a grounded UAS. Some platforms now won't fly if the software and firmware are older versions that require frequent updates. Updating is a whole lot easier than it use to be, but some models still require some basic computer skills to achieve this.

The software and application interface and user experience of today's commercial UAS platforms is feature rich and a real joy to operate. GPS, telemetry, battery and power management, range, video and photography access and control are all becoming commonplace. If the RF signal is interfered with or lost for any reason, the UAS will return-to-home (RTH) and land on its own, without instructions from the operator. Many other safety features are "burned" into the software and firmware so that the elation of flight is all the operator needs to be concerned with. Gone are the days of guesswork, stopwatches, and uncertainty about your platform status.

Post-Processing

Post-processing is the work done once the UAS has landed and the data captured during the flight is translated to consumable information for the intended audience. This phase can be as simple as removing an SD card from the UAS and copying the contents onto the hard drive of a computer. Conversely, the process could involve a labor-intensive editing process before it can be analyzed or packaged into a format suitable for release. For example, the photographs taken during a flight might need to be georeferenced and stitched together, then overlaid or integrated with other known data or imagery.

Let's look at a simple example. A mining company needs to determine the slope angles of a strip mine in order to maintain a safe working environment. A slope that is too steep is susceptible to shifting or collapsing, putting people and equipment in grave danger. In order to measure the slope, the mining company wants to use a UAS with a high precision (survey grade accuracy) imaging camera. The UAS operator creates the necessary flight plan, executes the plan, and then provides the imaging data to IT department for processing. IT takes the raw flight imaging data and runs the data through specialized software to determine exact slope angles from the UAS flight data. Then it compares this with any previous flight data to determine if any changes are measurable. The mining company also has ground composition data obtained from the U.S. Geological Survey (USGS) and this data will be overlaid to show mineral content and density.

To collect even more information the mining company wants a 3-dimensional (3D) model of the mine so the slope angles can be viewed in any axis. This is done today using specific flight plans, which provide suitable coverage of the area of interest with software that is specially designed to assemble images into a 3D model. This undertaking consumes huge amounts of computer processing and often requires dedicated computer resources to process the large datasets.

Post-processing has become so specialized and resource-intensive that several companies have been created with the sole mission of performing post-processing aerial content or imagery. They have several data formats they can handle and often have turn-around times of 24 hours or less. Post-processing can get complicated and time and resource intensive quickly.

UAS INDUSTRY MATTERS | PART 2

TECHNOLOGY MOSAIC

Brave New World

The wonderful world of drones is progressing at a blistering pace. Technological advances are improving the safety of aerial platforms. The regulatory environment is rapidly adapting to expanding demands of this new industry. Privacy concerns are ever present, but pessimism is giving way to reason and understanding. All of this is good, but are we there yet? Nope, not by a long shot.

I believe we are still in the early stages of this exciting industry and many opportunities are present and on the horizon. If we use Introduction, Growth, Maturity and Decline as the four phases to describe the commercial UAS industry we are, arguably, well into the Introduction phase of industry development. We see many entrants providing platforms across all the various types, whether it is a fixed-wing, rotary-wing or a multi-rotor. The emergence of defacto standards is occurring at regulatory and individual levels.

The convergence of technology from military applications, user experience, software, and firmware has rapidly evolved leading to advances that improve the usability and safety in and across the UAS ecosystem.

As this pace continues and normalizes with the continued adoption and R&D of these platforms, new and exciting applications and capabilities will unfold and emerge in the near-term.

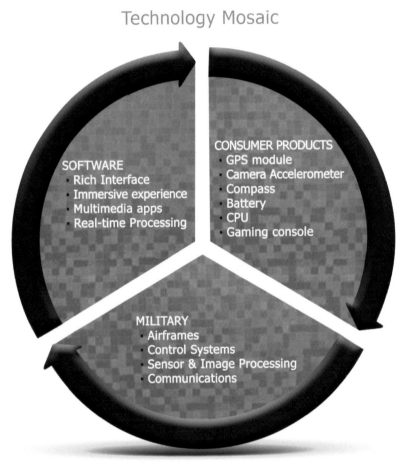

Figure 13: Technology Mosaic

The rapid advances made in consumer products, particularly miniaturization, software, and firmware will continue to increase capability and safety of the commercial UAS industry. The current velocity of innovation will spur new vertical and horizontal markets and open new opportunities and markets globally.

Four integration approaches have developed in the commercial market over the past year or so: controlled integration Apple-centric, open source, Linux or Android-centric, and finally proprietary. All approaches are currently viable and market demand will continue to spur innovation between each of these approaches over the next several years. If we look at the tech trends of the past 20 years, I believe we will see market behavior similar to the wireless/mobile, IT software & infrastructure, and .com periods.

One thing is certain, the future of UAS technologies and their benefits are only limited by one's imagination. Imagine using a commercial off-the-shelf UAS costing $1500 to fly into Fukushima immediately following the reactor meltdown – how valuable would that be? Imagine the valuable information gathered using a small fleet of inexpensive, rapidly deployable drones with thermal, infrared, LiDar, multispectral instruments immediately following the Deepwater Horizon event, or a train derailment, or hurricane Sandy, or a tornado. There is a lot more drones can do besides delivering your mail or packages. The technology to do all of these things safely exists today and will only improve in the coming months and years.

Next imagine using a fleet of low-cost drones used to monitor and manage a large agricultural facility. Using available technologies, a farmer could reduce his use of fertilizers, irrigation, and lessen some of the growing guesswork associated with crops. By using technologies like IR, multispectral imagery, 3D modeling, a farmer could maximize yields while reducing the negative environmental impact of excessive fertilizer and irrigation. This is an obvious win/win for everyone!

The US policy on UAS is rapidly evolving as well. The US UAS market is the most attractive in the world. The US has the greatest market influence in the world to drive demand, innovation, R&D, and resources to drive this capability into the future. The FAA, to their credit, has been very responsive to the market demands of this new industry. It was only 12 months ago that the government was strongly opposed to commercial UAS use. Now the FAA is "streamlining" their process and adapting to "enable" the responsible use of UAS technologies in the USA. They are acting prudently and responsibly in their support of this new industry.

Where do we go from here? It is fascinating and fun to guess. Autonomous UASs are already coming on-line, coordinated autonomous drones are already a reality, beyond line-of-sight is already available, and we are still in the infancy of this new industry. I have never been more excited about the benefits and possibilities of humanity and technology as I am today. Do we have issues/challenges before us? Of course, we do! Will that ever change? No. Can we use these technologies to the benefit of society without compromising safety, securing and privacy? Unequivocally, YES!

Where do we go from here? We are only limited by our imagination!

Convergence of Technologies

USER INTERFACE (UI)

The user interface (UI) of today's UAS platforms has seen many changes in 2015 alone. First and second generation platforms had little or no user interface to speak of. You had to connect the platform to your computer and use buggy software to configure the firmware in your UAS. The constant tweaking, upgrading and configuring led to many control issues, since many of the people making the changes had little understanding of what they were actually changing and how it might impact other aspects of flight control. We all would vigorously search the Internet for useful information or ideas on what to do in order to fix or reconfigure a firmware parameter. Looking back at the many issues I experienced in the early days of my UAS flying experience, I can attribute 99% of the problems I experienced to user error. I either screwed up a configuration change, I didn't properly calibrate one of the components or modules, or I did it in a sequence that created other problems. I am thankful that major platform providers through their UI have remedied all of these types of issues.

Today's off-the-shelf platforms have wonderful interfaces that expose a huge amount of invaluable information to the operator or user through an elegant application that runs on a mobile device (phone or tablet) or laptop or another computer. The app gives precise information pertaining to satellite navigation, battery health, RF and video strength, camera configuration options, a moving map showing where your UAS is and in what orientation, telemetry data (altitude, airspeed, orientation), proximity to restricted airspace, and a whole host of other necessary information for the safe operation of a UAS. Current generation platforms are light-years ahead of where previous generations were. I am sure the features and functions that enable safe operation will continue to advance. Every previous advance has truly revolutionized this industry, and will continue to evolve in amazing ways. So many technological advances shift the heavy workload and guesswork from the operator to software and firmware. This allows the operator to focus solely on the task at hand, whether it is flying for fun or commercial purposes.

Technology vs. Functionality Obsolescence

All these advances have also resulted in technology and functional obsolescence. Although I still have my first generation platforms and they still fly and are as functional as they day I bought them, they are technologically obsolete because I have become so dependent on the information available on current generation platforms. My latest platform uses multiple satellite navigation technologies [GPS (U.S. based) and GLONASS (Russian based)] to permit a level of positioning accuracy, coverage, and reliability that improve the safety of each flight.

Battery strength indicators remove the guesswork from the UAS platform endurance. Some even have a gauge that shows how much time remains until you reach pre-determined levels (safety margins) or how much power is needed to return to home. Finally, when power reserves are desperately low, if no operator commands are provided, the UAS will automatically initiate the auto landing sequence. Other advances are improving the safe operations and preventing accidental or unintentional reckless operations. For example, many new platforms are preconfigured to stay below certain altitudes and they cannot be operated within a certain distance from airports. These and other built-in safety features go a long way to reduce the risk to people and property.

I like this trend though I understand that operations above the limited altitudes may be necessary and operations near an airport may be required for various valid reasons. Managing the exceptions of the preconfigured limitations is doable. I have heard people express privacy violation concerns when a manufacturer builds into its product the ability to remotely limit or terminate its use. These are separate matters and also warrant careful consideration and debate.

REGULATORY ENVIRONMENT

US Regulations

The U.S. regulatory environment is rapidly evolving to meet the increasing demand for airspace suitable for commercial UAS use. Much debate and deliberation has occurred to discuss safety, security, and privacy issues associated with the commercial and private use of any UAS. The issues seem complex and varied in many regards, yet simple and straightforward in others. One thing everyone can agree on is, any UAS conduct that is reckless, malicious, or in violation of individual privacy in any form must have consequences commensurate with the infraction.

Some of the best regulatory policies in the US have focused on a risk-based framework with compliance requirements based on the level of risk associated with the flight operations. For example, if you wanted to fly a UAS in a desolate location (middle of the desert or some other remote area) then little or no regulatory restrictions apply. Conversely, if you wanted to fly over a densely populated city (Washington, D.C., NY, Seattle) then the regulatory requirements would be much more restrictive and onerous. The U.S. has been swift to respond to the needs of the UAS industry. Although some would argue that they haven't gone far enough or moved fast enough with any regulatory policy changes. If we look at all activities in the U.S. airspace – commercial aviation, general aviation, skydiving, hang gliding, parasailing, sport aviation, hot air balloons, military aviation, crop dusting, weather balloons, blimps and many others – all must conform to some basic understanding of how to operate in our skies without interfering with other possible uses. In any/all weather conditions, the issues, complexities, and risks become more clear to see and consider. As a pilot of 20+ yeas I have a great respect for all the facets of our airspace and appreciate the safety that is built into the system.

Integrating something new into our airspace may seem easy if you reside in an area with wide-open spaces that sees little or no air traffic. As you get closer to metropolitan areas the congestion and risk considerations increase exponentially. Also, when you look at the pervasive use of recreational UAS use, the risks go up even more since most recreational / hobby users have little understanding or experience with the U.S. national airspace system. The UAS platforms that anyone can purchase online or from a hobby shop have many of the capabilities of their larger brethren used for commercial or even military purposes. As the numbers of commercial and recreational use increases, so does the probability of a catastrophic accident occurring. We all agree we don't want to see something like this occur.

FAA Categories

FAA IDENTIFIES 3 TYPES OF OPERATIONS	
PUBLIC OPERATIONS	Governmental
CIVIL (COMMERCIAL) OPERATIONS	Non-Governmental
MODEL AIRCRAFT	Hobby /Recreation only

Figure 14: FAA Categories

PUBLIC (GOVERNMENTAL)

Public Aircraft Operations are limited by federal statute to certain government operations within U.S. airspace. Whether an operation qualifies as a public aircraft operation is determined on a flight-by-flight basis, under the terms of the federal statute. The considerations when making this determination are aircraft ownership, the operator, the purpose of the flight, and the persons on board the aircraft.

CIVIL (COMMERCIAL)

Any operation that does not meet the statutory criteria for a public aircraft operation is considered a civil aircraft operation and must be conducted in accordance with all FAA regulations applicable to the operation. This is where commercial UAS/drone use falls.

MODEL (RECREATIONAL/HOBBY)

The statutory parameters of a model aircraft operation are outlined in the FAA Modernization and Reform Act of 2012. Individuals who fly within the scope of these parameters do not require permission to operate their UAS; any flight outside these parameters (including any non-hobby, non-recreational operation) requires FAA authorization. For example, using a UAS to take photos for your personal use is recreational; using the same device to take photographs or videos for compensation or sale to another individual would be considered a non-recreational operation. The lines between recreational/hobby use has been a hot topic for the past couple of years. In some cases, hobbyists who've posted their videos on the Internet and monetized their video content have found themselves classified in the commercial category due to the money they received for people viewing their Internet content.

Exemption & COA Process

As we rapidly move towards the FAA's finalization of a new body of rules for small UASs, as spelled out in their Notice of Proposed Rulemaking

(NPRM), we must all prepare for a new aviation and airspace reality.

Figure 15: Floodgates

The current approval by exemption process utilized by the FAA to authorize unmanned aerial operations for commercial and public sector purposes is not a sustainable solution for the UAS industry. The FAA is already overwhelmed trying to review and manage the more than 2500 Section 333 Exemption requests flooding the system. The influx will continue and the current trend shows the number of requests outpacing approvals – and the margins are increasing! The exemption process is onerous and cumbersome, yet it currently does and will continue to have a place in the UAS industry going forward. Will the NPRM Help? Let's take a look.

The FAA's NPRM[1] is a good framework and is a step in the right direction. Everyone associated with the industry agrees something needs to change and the sooner the better (so long as it is easier for everyone). There is a lot to consider when permitting a UAS of any kind to

---

1. See more about the FAA's new rule set her: http://www.faa.gov/regulations_policies/rulemaking/media/021515_sUAS_Summary.pdf

operate in the US National Airspace System (NAS) and we are going to focus only on a few aspects of the proposed NPRM for the sake of brevity in this book. Today I want to assume the NPRM is adopted and implemented as written. When this occurs (in mid/late 2016) I believe the UAS industry will forever be different. The current process reminds me of the pace car at an Indy 500 race with the pace car representing the current regulatory process. Imagine the pace car is coming off the final turn and about to drop into the pit row. Then all the race cars explode and begin jockeying for position. I believe this is a great analogy for the new proposed rules. Once they are finalized, look out!

Will the new rules do away with the need for the Section 333 Exemption? NO! Will fewer companies and individuals pursue the 333 Exemption in the future? Absolutely! Why? Well, looking at the research presented in the Association for Unmanned Vehicle Systems International (AUVSI) "first 1000 exemptions" we find that a significant majority of the current exemptions are for real estate and general photography purposes (which combined represent greater than 50% of the first 1000 exemptions). Once the new rules are in effect, many of these same exemption holders and, I believe, the vast majority of new requests will opt-in to the new and less burdensome rules. The operator requirements under these new rules are so much lower than the Section 333 Exemption process that this will be a no-brainer.

Figure 16 provides a few examples of the new rules that have been proposed:

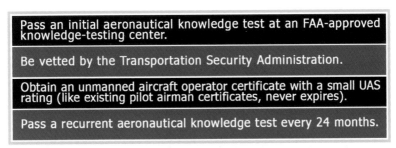

Pass an initial aeronautical knowledge test at an FAA-approved knowledge-testing center.
Be vetted by the Transportation Security Administration.
Obtain an unmanned aircraft operator certificate with a small UAS rating (like existing pilot airman certificates, never expires).
Pass a recurrent aeronautical knowledge test every 24 months.

Figure 16: FAA Rules

These are far easier to satisfy than the current licensed pilot, body of knowledge, and medical requirements, and much less restrictive with regards to platform registration and reporting requirements.

So these new rules will stimulate a flurry of activity. As an example, once the new rules are in place, every realtor (1M+) in the country could obtain permission (provided they can meet these new reduced standards) to conduct their own aerial photography or videography work. Is this a good thing? What do you think? I have no objections to safe and responsible operations of UASs. But I can say with confidence that obtaining a high degree of proficiency in flying a UAS is no trivial task, which seems to be a popular misconception.

More than 20 years ago it took me a year and about 50 flight hours to get my private pilot's license. At that point, I thought I was ready to fly anywhere and anytime. It didn't take me long at all to realize how little I actually knew about flying and what was really involved or at stake. It wasn't until after obtaining many more hours and additional certifications that I realized how complicated and risky it can be. Obtaining a private pilot's license demonstrates basic flight proficiency, but it takes a lot of time and operational experience to really be a good pilot, and it has little or nothing to do with a piece of paper.

If we look at realtors (which is just one small subset of the UAS industry) as an example and say that just 10% of the 1M+ realtors pursue a drone for commercial purposes, then that is 100,000+ new operators and platforms flying in the air around the country. If just 1% of these operators have a reportable "incident" this represents 1000 incidents that may require FAA and NTSB investigations, which could further stress the current regulatory framework. Once an aircraft has a reportable accident or incident the FAA and NTSB are quite often pulled in.

Earlier we said the FAA Section 333 Exemption requests (2500 and counting) are outpacing approvals and this was not a sustainable process. Once the new rules take effect I believe the number of requests to operate under these rules will increase 3-10 fold in just the first year alone, and will accelerate in subsequent years.

Each one of these requests will need to comply with the requirements listed above as well as all other requirements under the NPRM. If we stick with our realtor example, then 100,000 operators will need to demonstrate flight proficiency with their drone of choice – whether it is a fixed-wing, rotary-wing, multi-rotor, or one of the other types – to an FAA certifying entity. Now things get quite interesting.

Now, back to me becoming a pilot. When I was learning to fly, I had a flight instructor, who when he thought I was good enough, would have me fly with the chief flight instructor. When the chief thought I was ready we would schedule a "check-ride" with an FAA examiner, who would then test my flight knowledge and flight proficiency. This process took a lot of time and coordinating with an FAA examiner often took weeks to schedule. Back in those days, pilots learned in two primary planes (a Cessna or a Piper).

If we look at all the different kinds of drones in the marketplace today, our heads quickly spin with all the options. Now extrapolate this to the need to have FAA examiners (or designates) who have the proficiency to in turn certify your proficiency. We get to some rather large numbers rather fast. I don't know how many UAS examiners the FAA currently has, but I could offer a guess and say, not nearly enough.

Having been in the UAS industry as long as I have, I can say with confidence that currently there is a scarcity of qualified drone operators. Moreover, I will take it a step further and say there is a shortage of operator candidates as well. Yes, many schools (aviation oriented) and the military are adding good numbers of potential pilots to the marketplace (and this a good pool of candidates). But flying a plane and flying a drone is not the same skill set! Good training is even harder to find. I am often asked where we find qualified operators. My response is simple – we don't often find them, we have to "build" them by training them.

Opening the floodgates with the finalization of the NPRM is exciting, and a thriving UAS industry is good for the country on so many levels! However, lowering the 333 Exemption bar does not solve all the problems; in fact, it creates many other problems. Additionally, it shifts the regulatory burden and risks further downstream – certification, oversight, enforcement, etc. For now, I remain encouraged and excited about the future of the UAS industry and optimistic the industry will quickly adapt and improve with the demands of the markets it serves. We just need to be smart about managing hyper-growth and find a way to keep our heads above water.

US Airspace

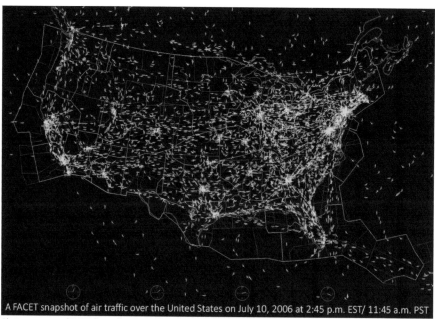

A FACET snapshot of air traffic over the United States on July 10, 2006 at 2:45 p.m. EST/ 11:45 a.m. PST
Photo courtesy of NASA (http://www.nasa.gov/images/content/535942main_facet1.jpeg)

Figure 17: Congested Airspace

The FAA regulates the U.S. airspace; it is their responsibility to establish policies that govern the use of our airspace. This is a massive undertaking when we consider that our airspace is used for recreation (i.e. skydiving/hang gliding), commercial, industrial, military, and space exploration. Any and all activities that want to utilize our airspace must comply with FAA regulations. Flying a small remote-controlled (R/C) hobby airplane in an open area may not seem like a big deal in remote regions unless it is next to an airport or some other controlled area. As is evident in Figure 17, the airspace in the United States alone is incredibly congested and potentially hazardous to the unaware.

If you are unfamiliar with airspace restrictions and all the procedures associated with operating in, near, or outside various airspaces you are not alone. Pilots go through an extensive education program to learn about these restrictions and procedures. The two categories of airspace are regulatory and non-regulatory. Within these two categories, there are four types: controlled, uncontrolled, special use, and other airspace. As an example, Figure 18 presents the five airspace classifications (A-E) that make up the controlled airspace. Uncontrolled, special use, and other airspace classifications also have procedures for operation with them. Our airspace system serves many uses. The intent here is to introduce the complexity associated with operat-

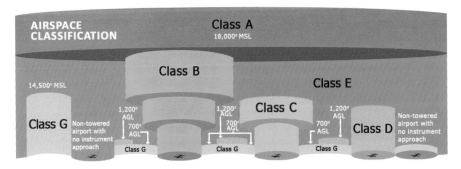

Figure 18: AirspaceClassification

ing within our airspace system. While a large portion of pilot training is focused on flying an aircraft, the body of knowledge that goes along with being able to safely operate an aircraft can take much longer to acquire and comprehend.

MARKET CONDITIONS

Global

The global UAS market is hot and growing at impressive rates in virtually all areas of the world. As the demand continues to grow and technological advances, reliability, and safe operations continue to improve, several existing manned aerial activities will be replaced by unmanned aerial platforms due to the lower cost associated with the generated content. More than replacing existing manned aerial mapping, photography, LiDar, and other aerial sensing data-capturing activities, new applications will drive demand.

We have already discussed many applications that have immediate benefit and value from using these capabilities, and I believe this is just the beginning of what will surely be a revolutionary impact in all markets and segments. The benefits already outweigh risk or liability in many areas, and as technology evolves in this industry so will new applications and opportunities.

Other Regulatory Environments vs. US

Around the world, countries are already beyond initial regulatory policies. The U.S. has been lagging behind several other developed and developing countries in having a regulatory framework for commercial UAS use. That said the U.S. market has already begun to leverage what is working in other regulatory environments and how to avoid introducing unnecessary friction into an industry still in its infancy. The Federal Aviation Administration (FAA) has been quickly adapting to the market needs while ensuring they are adhering to their charter of providing safe airspace for everyone that flies within it.

Other countries have risk-based regulatory positions which permit greater latitude based on the inherent risk associated with UAS operations. For example, if you are flying a drone in a remote unpopulated area with little or no danger to people or property, then the regulatory standard is more lax and requires lower compliance standards. As the risk increases so does the compliance requirement. For example, say you wanted to fly a UAS over a densely populated urban area to provide aerial content for a movie or commercial / industrial purposes (photos, video, LiDar, 3D mapping, etc.) then the regulatory standard would be the most demanding (must obtain pre-approval, be certified, have approved equipment and operating procedures, etc.). All great common sense regulatory positions and they also are more enabling postures for the industry.

Growth

All analysis and estimates point to massive growth in the commercial UAS industry in all areas and developed markets. These estimates don't always tell the whole story and it would be difficult for them to do so, since the industry is moving so fast and moving into so many aspects of industry and life (package delivery, mining, mapping, energy, real estate, disaster & humanitarian response, and many more).

Plus these new technologies are pulling from and expanding and supporting existing industry as well. The most popular platforms today use electric propulsion, which is driving even more advances in battery innovation. These new devices are powerhouses of processing instructions and they are doing more and more with every release of a new product. This places high demands on power and power management processing, which has drawn the attention of chip manufacturers. Intel announced a move into this space and has already been putting R&D effort into sense and avoid capabilities.

Applications for every aspect of drones are exploding, the "there's an app for that" is taking hold already in this industry. So are selfie drones, which seem to appeal to so many sport adventurers in many areas. The future is bright and exciting for this industry.

The U.S. market is already exploding at unprecedented rates, so much so the government has serious concerns about the projections for the 2015 holiday season – which projects the addition of nearly one million new drones to its airspace (all these will be for recreational / personal use). With these kinds of numbers, even the most conservative estimates point to an increase in the odds that more and potentially severe incidents will occur.

Drone / UAS / UAV Providers

Originally this section was going to tiptoe into UAV/UAS providers, but with so many existing and new providers entering the market the list would be too large to cover them all. And I imagine that by the time you read this book a new entrant will emerge or be making announcements. What can be said with confidence is that the industry will initially support a great many providers and niche players. Though at some point, likely many years in the future, some level of consolidation will occur.

The one thing that hasn't happened yet, but will at some point, is the entry of a well-established player from the military sector entering the commercial space. It is only a matter of time. Additionally, it wouldn't be a stretch to see something from a Boeing or Airbus once the industry shows more maturity. All very exciting outcomes to ponder and consider; all bets are still on the table.

Career Opportunities

We can't talk about all these great things without talking about "WIIFM" (What's In It For Me) or career opportunities. Careers in the industry are already available, though they may not be what everyone gets excited about. When I'm conducting seminars or training sessions, or just out and about, everyone inquires about flying a drone because that's what they want to do. It's cool, it's sexy, and for sure it is a lot of fun. In reality, the pilot or operator is only one aspect of an increasingly complex ecosystem.

If you look for an example of what it takes to operate a drone, you could look at a military drone operation to get an idea of what is involved, even if you are only doing it for recreational purpose or for basic commercial purposes (i.e. real estate photography). What you will discover is you need an operator, a sensor / payload operator (even if it's just a POV camera), a spotter, navigator, maintenance staff, data processing (imagery analysis), application / software engineers, and many more roles. For the hobbyist or recreational user these roles may be a single person or a child and his/her parent. For all drones out there all these functions are still necessary, even for an off-the-shelf recreation UAV. All of them require operational proficiency, maintenance, and some sort of post processing. Maintenance isn't just in the form of motors and mechanical needs, it is increasing in the form of firmware & software maintenance or patches. All platforms I am aware of require some level of understanding in all of the areas mentioned. Technological advances are making it easier all the time to perform these functions, and they are increasingly safer since many platforms today have black boxes. Yep, if you crash, then you can go to the black box and find out why and what happened – user error or system malfunction. They also have an increasing amount of data about the health and performance of all critical functions.

For the U.S. commercial industry as of now only an FAA licensed pilot can operate a UAS/UAV/drone for commercial purposes in the U.S. airspace, and only if they have approval to do so (remember the Section 333 Exemption and COA discussed above).

EDUCATION & TRAINING

This is an area that is already in high demand, though it is also an area that is wildly unorganized, unstructured and filled with well-intentioned programs that often mislead students looking to enter the commercial UAS space. It is an unfortunate reality of an industry that is still in its infancy.

Colleges and Universities are beginning to offer curriculums that support the industry, and this will continue as the market matures. Though with the velocity at which the market is moving, engineering and software development types of skills are likely to remain hotbeds for those looking to enter this new space. The operator/pilot role is going to be the long-pole in the tent until the FAA pilot certification is relaxed, simply because any FAA pilot certificate / license is time & money intensive. The current standard is a sport pilot's license, which only requires 20 hours of flight time. These 20 hours often take four to six months or more and thousands of dollars to obtain. For existing pilots, not all are willing to risk putting their hard-earned license at risk to fly a commercial UAS.

CERTIFICATION

In addition to body-of-knowledge education and training, flight proficiency is an area that is rather subjective at the moment. Many folks claim to be great UAS / drone pilots or operators. Everywhere I go I am asked, "Where do you find qualified operators?" I am surprised myself that so few exist. Yes, there are people who can fly a UAS, but once you start flying in a strictly controlled manner and perhaps in an area with little margin for error then this is quickly where deficiencies become great liabilities. I see it over and over. I see videos where people are flying great in open areas and take amazing pictures and videos. However, many commercial applications don't offer the luxury of open spaces. They are often plagued with obstructions and are in areas that may have consequences should there be any issues — real or perceived. You don't always have to crash to have a major issue or problem when flying commercially. Commercial standards and expectations are always high when money is exchanged for services.

Certification beyond what the FAA requires for commercial purposes is highly subjective and often specialized (aerial photography or cinematography — both very different). A combination of body-of-knowledge and hands-on proficiency that covers the systems, controls, flight characteristics, safe operating procedures, emergency procedures, sensor / payload operation, maintenance, post-processing, and documentation should be, at a minimum, covered in sufficient detail to allow safe operation.

A well-structured program will have multiple tiers/levels that build in both body of knowledge and proficiency that have entry/exit requirements for each level or tier. For example, a comprehensive program may take weeks or months to complete due to proficiency standards, not time standards. Like any FAA pilot program, the number of hours is deceiving. The sport pilot license, which calls for a minimum of 20 hours, cannot be obtained in a single 20-hourcontiguous session.

The 20 hours represents a minimum level of proficiency that takes a prolonged duration to obtain, usually one flight hour at a time. One (1) flight hour is usually achieved in a single two-hour instructional session. Each instructional session usually involves several hours of preparation (usually body of knowledge related and entails a lot of reading). Similar proficiency standards for UAS certifications are difficult if not impossible to find.

APPLICATIONS

How Drones are Used

Ok, we have made it this far. Now that we have a deeper understanding of what drones are and how they operate, let's look more closely at how they can be used in the real world. This section alone could represent volumes of books about specialized applications using this type or that type of drone. For our discussion, we are going to keep to topics that most everyone can relate to or understand. I will preface this with saying we are truly only limited by our imaginations when it comes to how a UAS may be used. Here are few examples I will offer.

First Responders

First Responders have a unique opportunity to leverage the capabilities of UASs in responding to catastrophic events. Today we are going to explore how they may be used for Hazardous Materials (HAZMAT) incident responses. Our examples are hypothetical, but the capabilities and features are real and available today in many of the commercial off-the-shelf platforms that are available today. High definition video is standard on the majority of available aerial platforms. Adding additional payloads is possible as well, adding a thermal, Infrared (IR) or even multispectral sensors. Adding any one of these capabilities to the first responders' "toolbox" makes sense, saves lives and reduces exposure to high-risk environments for our first responders.

Figure 19: Train Derailment

The railway system in the United States is a vital component of our infrastructure; it is aging, in need of repair, and is at or above capacity in many areas of the nation. Add to that an increasing demand for rail transport, increasing the number of hazardous materials being moved by rail, and the probability of serious accidents becomes more and more unavoidable. On top of that, first responders in most municipalities are faced with year over year budgetary cutbacks across the board. When accidents happen, response time, situational awareness, and incident containment are vital.

For our first hypothetical scenario – train derailment – let's assume the following. The train is 150 cars, containing 20% HAZMAT (randomized amongst the cars), traveling within 50 miles of a major city. The derailment occurs in a semi-rural county, first responders have a small HAZMAT team trained in this scenario, the incident takes place at 6am, the derailment occurred at moderate speed on a curve, 30% of the cars have left the track and some are leaking, fire risk is high, and the accident happened in a difficult to access area. Finally, the prevailing weather is clear, dry, with a light 5-10kt winds blowing in the direction of densely populated areas.

First responders (local police and fire) arrive on the scene by 6:20. The HAZMAT team arrives and begins assessing their approach at 6:30. The train's manifest indicates several highly dangerous & flammable chemicals are among the 150 cars. The on-scene commander has already requested additional support from surrounding counties, the NTSB has been notified and is preparing a response team.

The on-scene HAZMAT team has two (2) Level A suits (one for primary crew members and one for backups) certified for the chemicals and fire hazards. This configuration would only permit a limited amount of time for the damage assessment and little or no time for response and containment. Additional help would be needed. Response time and containment time are crucial when dealing with a chemical incident. The hazards posed by the chemicals themselves are serious and even fatal. When combined with other chemicals, the potential chemical reaction may lead to a significant event forcing the evacuation of large population areas and the cleanup costs to soar into the millions.

In this scenario, rapidly being able to assess the scene and determine the level of leakage or environmental exposure is crucial. Minutes count. Having a small commercial-off-the-shelf (COTS) UAS that costs less than $5000 would allow the team to quickly assess and record the entire scene and deploy the available resources to the most critical areas until a larger response team arrives. Utilizing additional sensors (IR or thermal) would allow a more detailed assessment to be made without taking more time while limiting the exposure of people to the high risk/hazard environment. Additionally, with any Internet connection (cellphone, Wi-Fi or otherwise) the video signal could be sent in real time to additional responding teams, allowing them to formulate an action plan for when they arrive, saving precious time and allowing them to go into action as soon as they arrive on scene.

Today HAZMAT teams are highly skilled first responders that are essential in many scenarios. Adding this type of capability to their toolbox is a no-brainer; it saves lives, increases response times, coordination between agencies, and reduces environmental impacts. The benefits cannot be ignored and the entry price is affordable for any Fire Department.

Deepwater Horizon from the View of an Unmanned Aerial System (UAS)

BACKGROUND

The Deepwater Horizon oil spill began on April 20th, 2010 in the Gulf of Mexico on the BP-owned Trans ocean-operated Macondo Prospect. Eleven people were never found, and it is considered the largest accidental marine oil spill in the history of the petroleum industry, an estimated 8% to 31% larger in volume than the previously largest, the Ixtoc I oil spill. Following the explosion and sinking of the Deepwater Horizon oil rig, a sea-floor oil gusher flowed for 87 days until it was capped on July 15th, 2010. The US Government estimated the total discharge at 4.9 million barrels. After several failed efforts to con-

Platform supply vessels battle the blazing remnants of the off shore oil rig Deepwater Horizon. A Coast Guard MH-65C dolphin rescue helicopter and crew document the fire aboard the mobile offshore drilling unit Deepwater Horizon, while searching for survivors. Multiple Coast Guard helicopters, planes and cutters responded to rescue the Deepwater Horizon's 126 person crew.

Figure 20: Deepwater Horizon

tain the flow, the well was declared sealed on September 19th, 2010. Some reports indicate the well site continues to leak.

On April 20th, Aerial Strategies receives a frantic call from one of its premier customers who happens to have a large position in BP, and they are extremely concerned about the news reports about the Deepwater Horizon explosion. The customer, ACME Investments, asked Aerial Strategies to conduct an immediate risk assessment of the "real" situation on the ground. If new reports are accurate, BP will suffer a financial loss that they may never fully recover from, and ACME Investments may lose a significant portion of their energy portfolio. Below are some of the concerns that the executive team conveyed to Aerial Strategies:

ACME Investments executive team explains the conundrum and the looming decisions it faces with BP.
ACME Investments executive team identifies what they think will create the most near-term volatility in the BP stock price;
Bad PR/headline news from the front lines (e.g., video of seagulls stuck in muck, tar balls spread over the beach, oil slicks washing up on fishing boats, etc.) – could easily trim 1-2% off the stock price in a given day
Legal liability as a result of damage to fishing/oyster grounds, shrimpers, loss of vacationer traffic to the beach (hotel/beach rental operators), etc.

ACTION PLAN

Aerial Strategies immediately jumped into action. They sent 10 drone teams to the region to begin a week-long risk and impact assessment. The risk assessment response team formulated an action plan to establish a baseline of areas predicted to suffer the greatest impact along the gulf. As shown in Figure 21, the initial baseline would cover 100

Figure 21: Gulf Coast-100 mile of affected coastline

miles of coastline. Each deployment team would be responsible for covering 10 miles of the coastline using a combination of fixed-wing, rotary-wing, and multi-rotor drones from the Aerial Strategies fleet.

Each drone was equipped with a multi-spectrum sensor capable of creating high precision imagery in the visual, infrared, and thermal spectrums. Each team would fly three times a day —low tide, high tide, and mid-tide — for seven days. Once all the content has been captured and analyzed they would create 3D models of the impacted areas. Furthermore, other information would be added to the assessment to further enrich the data, adding additional layers of information.

For example, as demonstrated in Figure 22, by adding weather data (previous, current, forecasted, and 10-year historical trends) the Aerial Strategies Risk Assessment Team could enrich their analysis thereby increasing the confidence of their assessment. By adding additional data – geological, coastal terrain, soil composition, Gulf currents, marine life density, bird migratory patterns, environmentally sensitive ar-

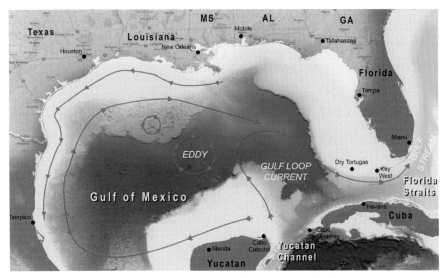

Figure 22: Gulf with additional weather data added

eas and a host of other data, the team could further determine the severity (and probable recovery time) of the impact on multiple levels (environmental, economic – fishing/tourism/shipping/etc. – social and other affected sectors).

IMPACT ANALYSIS AND FINDINGS

The outcome of the analysis was provided, on-time, to ACME Investments with an impact assessment of the BP event. The oil spill was going to be more significant than originally reported, environmental impacts would be less severe than predicted due to local ocean currents, wildlife impacts would be moderate but would recover quickly due to the soil composition of the marshlands, etc., etc. The biggest surprise in the assessment was not the Gulf coast economic impact, but rather the impact to the Mississippi river commercial barge sector. It would suffer a massive impact due to disruption in the traffic flow of the summer corn and wheat harvest.

ACTIONABLE INTELLIGENCE

During the week-long initial impact assessment, Aerial Strategies was able to provide incremental and live stream imagery and analysis to ACME Investments in real-time, providing their customers with actionable information during the entire assessment. ACME Investments was then able to manage their portfolio risk exposure daily. When seconds count and the difference between selling immediately or in five minutes when the next news reports splash across TV screens around the world, customers would certainly be glad they had the information to make the right call!

With the impact analysis and findings, ACME Investments was able to make adjustments to their portfolio to capitalize on the shift in market conditions on the Mississippi while minimizing their short/medium/long-term exposure to the situation in the Gulf. ACME Investments' worst fears of losing the majority of their BP position was never realized. Instead, they decided to short the barge industry on the Mississippi and then hedged their risk in the Gulf, without abandoning their long-term BP position. In contrast, without this knowledge, ACME Investments competition sold BP, got caught "long" on the Mississippi and lost several institutional investors who moved to ACME Investments.

Agriculture

This is an area that has immediate and significant benefits and value with respect to the use of drones. Many universities and companies are working vigorously on platforms, sensors, and applications specifically for this market sector. The benefits in agriculture are difficult to ignore and even harder to dispute. Japan and Australia, for example, have been using drones successfully for agricultural applications for years. Let's discuss a few short examples. Let's look at a large commercial or family operation that has 1000+ acres of tillable land under management.

First, the large commercial or family operation has over 1000 acres of tillable land they work. Let's say they get one or two seasons per calendar year of marketable product. Some farmers can get two or more seasons in any given year, depending on their agricultural zone and markets served. For simplicity, we can assume the acreage is used for general produce production. Lots of variables are factored into all aspects of fertilization, irrigation, and pest management. The benefits we will discuss will assume a relative comparison to keep it in perspective. We will assume all optimum conditions in an average or good year for rain, temperatures, pests, etc. Also, since we are not being specific about soil conditions or composition, climate zones, and several other factors, I am going to say for the sake of argument that 100 pounds of fertilizer will be used per acre and per season, and 100 gallons of pesticide per acre per season for our theoretical crop. One hundred pounds if anything is on the low side for an entire season for any crop. So let's agree to start there.

Using this, we would be using 200,000 lbs. of fertilizer and 200,000 lbs. of pesticide per calendar year for our 1000-acre operation (remember we have two seasons per calendar year at this operation). Let's also assume a cost of $.50 per pound for fertilizer and per gallon of pesticide, which would translate to $100,000 per calendar year for fertilizer and $100,000 for pesticide for a total of $200,000 / year for fertilizer and pest management. Irrigation is a tough number to assume, since it has so many variables all by itself, so let's set our irrigation cost/year at $0.00 for this discussion. Already, we can see we are talking some really big numbers and we already know these are in fact, much lower than real numbers, so the actual and real benefit will be greater than any benefit we can show here.

Finally, we have to assume some yield numbers for our crop. If our crop is a corn or veggie (bean, pepper, tomato, squash, etc.) then we can comfortably say we should be able to yield 100 bushels of any of these crops per acre / per season. Using this, we have 200,000 bushels of our crop per calendar year. Let's assume we can sell each bushel for $2.00 which translates to $400,000 / calendar year. Let's see where we can go from here using these numbers. We can also assume we have some other expenses to cover – temp labor, equipment, transportation, etc., etc. Let's assume another $100,000 for these expenses, though I realize they could be wildly higher or lower, depending on many variables. Using what we have so far, we are showing $100,000 of net income or profit. Let's invest $10,000 in a super fancy drone that will provide us real-time IR and multi-spectrum data that can be readily formatted for use in our farm equipment (this is already widely available in current market products, this is real and not imaginary, it's called tractor ready data). Finally, I am going to assume the operation is proficient and compliant in all aspects of UAS operations. Let's go flying.

Since this is our bird and we can fly it as frequently as we'd like, we will establish a grid pattern with GPS waypoints programmed into the drone. It will then fly the pre-programmed grid the same way every time all the time. Our datasets will then be easy to index, correlate and compare against other datasets. We will keep all data and use this data to establish historical reference points based on projected and actual usages (for fertilizer and pesticides). Also, we will be able to now specifically overlay yield per acre (perhaps even with greater resolution – think per foot) data over our grid datasets and compare it year over year. Great stuff so far right? ABSOLUTELY.

Since this is an established operation, we do know that we already have great information about our land, we probably have existing soil composition data (perhaps annually), moisture, and other important data. Our baseline flight is going to fly the grid to establish a baseline of soil moisture content. I feed this information into my planting equipment, which can then calculate the exact amount of fertilizer needed, per seed, per foot over the entire 1000 acres. All this information goes into our growing database for use later. We now plant.

Our flight program stipulates that we will fly twice a week to determine moisture content and temperature, and if necessary, we can fly on-demand or more frequently if conditions or data dictate. Now we are able to monitor every foot of our thousand acres using a UAS. With this data, we are able to immediately identify dry spots, high moisture areas, weed and pest density. Armed with this data, we are able to take the guesswork out of a crop management program. The operation would know exactly how the entire crop was doing (per square foot) and when and precisely where to apply fertilizer or pesticide. The amount of chemical waste (fertilizer or pesticide) is difficult to determine with precision. However, if we determine with higher certainty the exact amount needed to fertilize and when a more accurate estimate can be applied more to future demands. This would also apply and correspond to any/all pesticide or other chemicals used to manage a healthy crop. Any precision that can be applied to future expenditures has an immediate positive impact on the financials and potentially cash flow and profits.

For our hypothetical example, let's assume we can show a 3% reduction of fertilizer and pesticide consumption, this would immediately translate into a savings of $6000 (based on our $200K in chemical expenditures). It would also have the added benefit of less environmental impact and any negative soil burnout issues associated with chemical overexposure. This is a clear win-win and allows for better land management, stewardship, and responsible agriculture practices, which will carry forward for future generations. All good stuff, right?!

Now more good news. Let's assume that due to our better management practices – precise fertilization, irrigation, immediate and targeted pest management (loss reduction) – we are able to improve our yield. Let's assume we are able to increase our yield by the same 3% in the first year. Based on the $400K we used at the beginning we would improve our top and bottom line by $12,000 in the first year. Combined with our $6,000 chemical saving, we improved our business by a real and measurable $18,000. The $10k investment paid for itself in less than one year. The real number would also increase substantially due to reductions in equipment, fuel, and labor costs as well. Fuel and maintenance on a 1000-acre operation would also immediately show a reduction. If 100 hours of equipment time per machine were saved each year, the additional cost savings would be much, much higher.

If we extrapolate these numbers down to a smaller farm, I would expect the savings to be a relative comparison, though some big numbers in equipment, maintenance, and fuel may offer significant incentives for smaller operations.

By now I am hoping the benefits of these technologies are helping you realize how important these capabilities are to so many aspects of our economy and our lives, directly and indirectly!

Inspections

Inspections of major infrastructure (bridges, railways, towers, antennas, pipelines, cranes, water towers, power transmission lines, etc.) may offer even greater savings since these are generally areas that possess high risk for humans when compromised. Anytime we can reduce or remove people from high-risk activities we should be able to find a compelling cost/benefits analysis to justify the use of these technologies.

INTEGRATING UAS TECHNOLOGIES INTO YOUR BUSINESS

PART 3

BUSINESS ADOPTION & INTEGRATION

Aerial Platform (UAS / Drone) Enterprise Integration

By now nearly everyone has heard of, played with, or owns a personal drone. A year ago the mere mention of the word drone conjured up images of a military Predator sending a precision-guided bomb into a remote region of some far away place. They were not for enterprise consumption and even carried a stigma. A lot has changed with drone technologies in a year, and the public perception has changed considerably, too. The vast majority of people now understand that small Unmanned Aerial Systems (UASs) are very small and very different than their bigger ancestors they see being used by militaries around the world.

Today's commercial UAS reality is here. Imagine using a small UAS (like any of the 3DRobotics platforms or one from DJI, for example) to search for people after a catastrophic event (earthquakes, floods, snow storms, tornadoes, etc.). Imagine deploying one with the ability to pinpoint a leak of a poisonous gas emitting from a pipeline in a densely populated area, or to inspect a bridge truss using infrared, thermal, and high definition video – and beaming that information in real-time, to an army of experts and engineers around the world to best determine what fix is necessary. All of these are easily achievable with the technologies and capabilities available right now, at a cost in the range of $1,000-$5,000. The savings in economic impact, lives, and liability is huge!

Technologies from the military and the consumer products sector and the advancement in software development and engineering have converged in a way that permits these applications to become a reality.

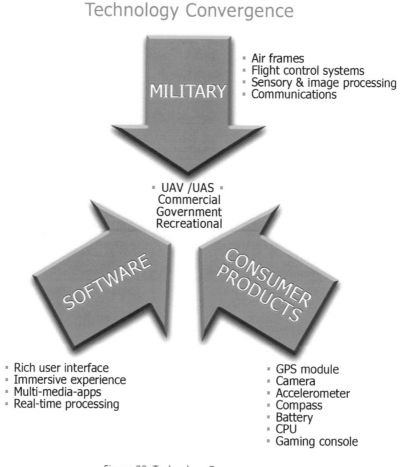

Figure 23: Technology Convergence

The military has been spending billions of dollars on research and development (R&D) and many of these technological capabilities have found their way directly into UASs for the commercial, public sector, and recreational markets – carbon fiber airframes, command and control systems and sensory technologies among them. Our cellphones are a marvel of the miniaturization trend prevalent in consumer products – they now have accelerometers, integrated GPS, high definition cameras, and many other amazing capabilities we only dreamed of a couple of decades ago. This miniature componentry has been creatively re-purposed for use in smaller UASs with size and weight constraints.

Finally, software has allowed us to say, "there's an app for that..." for just about everything. For aerial platforms, this has been a critical, recent development that has allowed much of the highly complicated flight activity to be handled by firmware and software, while leaving the operator / pilot to remain focused on the task at hand. The user interface (UI) to the aerial platform has now emerged as a rich experience with safety engineered directly into the aerial platform to reduce the margin for error. If the aerial platform loses communication with the operator / pilot, it will return to the place from which it took off. Some aerial platforms have no-fly zones built into the firmware of the platform, prohibiting it from entering restricted airspace (the White House, near airports, or other sensitive areas).

This is a very exciting time for this new industry; the benefits are astonishing and the liabilities / risks are becoming better understood by regulatory bodies, insurance companies, and businesses every day. The FAA has made huge strides in the past year to understand and accommodate small UASs into the US airspace. They continue to make prudent, necessary progress towards the commercial and public sector integration of these platforms into the US airspace. There is much progress still to be made, but we are quickly moving in the right direction. Now that we have this extensive and revolutionary capability, how do we assimilate and integrate it into our enterprise? Simply running out and acquiring a drone is appealing. However, the ecosystem and operating environment for these platforms are very complex. The complexity increases, sometimes exponentially, based on the application, the sensory payload, the operating environment, the post-processing requirements and the list goes on and on. This complexity can easily cloud the decision-making process and lead to sub-optimal results (as the example of the FBI acquiring a UAS fleet has demonstrated).But the benefits of prudent deployment justify the effort to get it right. Let's look at a very simple example. ACME Construction is building a concrete slab in the middle of nowhere for a developer. The concrete slab is going to support a data center for commercial customers within two years. The slab is going to be 100,000 square feet and will meet or exceed all building codes for the local region. The region is within five miles of a seasonal water tributary that feeds a waterway that eventually leads into a major river before heading to the open ocean. The construction zone is also within 20 miles of migrating geese that often use the waterways during their migration. So environmental concerns are high and the risk of despoiling the environment is substantial.

Before construction begins, ACME contracts with an aerial imaging company to establish a baseline of the construction area. This baseline is comprised of video, high-resolution still photographs, infrared (IR) and multispectral imaging to show the soil composition, terrain analysis, and any and all environmentally sensitive areas surrounding the construction site. Additionally, all available wildlife migratory patterns and environmental GIS information are overlaid with the aerial imaging to show compliance with contract and insurance requirements.

ACME then conducts several aerial surveys using unmanned aerial platforms, at regular intervals throughout the construction process.

All imaging is analyzed and compared against previously captured imaging data to identify any anomalies or compliance violations, and to limit future liability claims. The cost of this extra effort is a small fraction equal to about 1% of the total construction cost, and will pay for itself with the reduced insurance premium required of the construction project.

Over the 18-month construction phase, the aerial platforms captured three terabytes of imaging and sensory data. All of this data was assimilated, analyzed, stored, and processed into consumable and digestible information that would be disseminated internally and externally for private and public consumption. None of this data contained sensitive information and all of it was available through other public means. Effective data management policies and procedures were defined and implemented to support the retention of this data trove to meet contractual requirements and minimize legal liability for years into the future.

The project was a resounding success and was delivered on time and within budgetary parameters. ACME was able to complete the new data center, which has a continuing and substantially positive impact on the local economy, without any environmental issues. They did such a great job, they became the standard by which all other concrete slab measures companies.

In this example ACME had outsourced the aerial services for this construction project, thus avoiding the additional cost associated with acquiring, maintaining and managing a fleet of drones. Not to mention the need to keep up with the endless software and firmware updates, standard operating procedures (for platforms and for the related business process workflows), and the training and proficiency requirements necessary to maintain regulatory compliance. This strategy allowed them to focus exclusively on their core business: concrete slabs.

The integration of drone content into the enterprise will require consideration on many levels – business processes will need to be revised, human resources (HR) will need to define new roles and training programs for these roles, corporate governance will need to account for regulatory changes, organizational risk, and insurance-related liability as well as safety, security, and privacy concerns. Information technology will need to support large unstructured data, new analytical tools that leverage and take advantage of these new data sets, and the GIS capabilities that will enable a higher level of actionable business intelligence that will drive the business. This is the dawn of a new era for the enterprise, and the "perspective" as we like to say, is very bright and exciting.

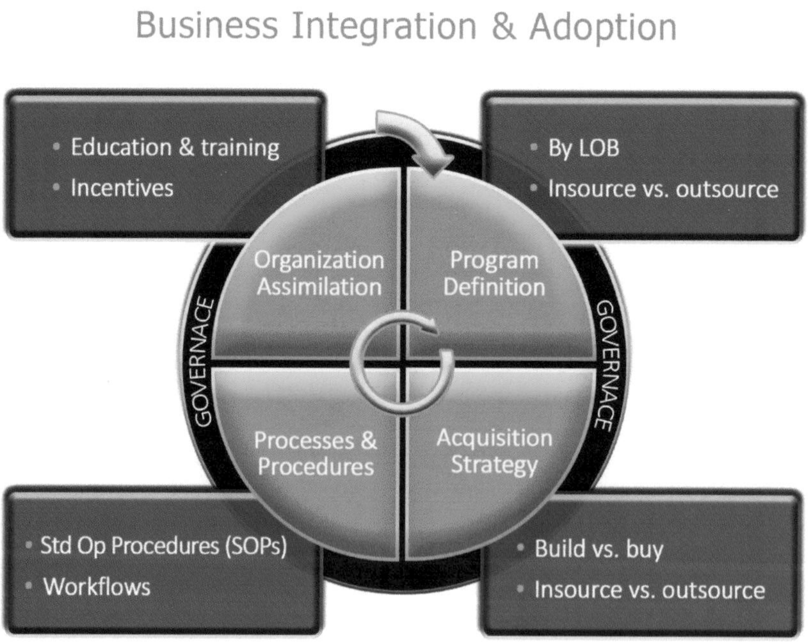

Business Integration & Adoption

- Education & training
- Incentives

- By LOB
- Insource vs. outsource

Organization Assimilation

Program Definition

GOVERNACE

GOVERNACE

Processes & Procedures

Acquisition Strategy

- Std Op Procedures (SOPs)
- Workflows

- Build vs. buy
- Insource vs. outsource

Figure 24: Business Integration & Adoption

Once an organization, be it commercial, industrial, or federal, decides to incorporate aerial content into their enterprise, many facets of acquisition, integration, consumption, and control need to be considered and potentially revised. Once the analysis has been completed and the value has clearly been articulated, an adoption / integration plan should be created. This plan will need to be comprehensive enough to manage, at sufficient level, so as not to introduce risk or liability issues into the organization. It would be easy and tempting for an organization to simply go online, order a seemingly suitable UAS and start using it. As we discussed earlier, the FBI and ATF attempted this to the tune of $2.3M with disastrous consequences. Don't be "that guy" or company.

Let's explore a few basic ideas a bit further, these are not comprehensive, since each organizational need is different and each of the topics we will discuss will need to be addressed at a level specific to the organization's or enterprise's needs and goals.

Insource

An organization or enterprise looking to get started may decide to do everything in-house and quickly run out to purchase a fleet of UASs to meet their identified need / requirement(s). Great. Insourcing has many advantages. But it may have some drawbacks if done hastily and without clear sponsorship, governance, and many other considerations. Advantages may, on the surface, appear to be low acquisition and operating costs when compared to outsourcing. A low Total Cost of Ownership (TCO) is always an attractive justification for "doing it ourselves" mentality.

Organizations attempting to start with an insourcing approach would be wise to ease into their program deliberately and incrementally. When speaking with organizations about a UAS strategy, I often use the analogy that purchasing any UAS is akin to purchasing a Gulfstream IV (corporate jet) for the organization. Since a UAS falls under the FAA purview (in the U.S.) it has all the same requirements a fleet of aircraft has: they must be rigorously maintained with documentation, qualified personnel must operate them, and all the regulatory policies, operating procedures, Information Technology (IT) infrastructure, and on and on must be in place "day one" of their use. Any organization considering a drone program would be well advised to obtain knowledgeable outside assistance in forming an adoption plan/strategy.

A well-defined plan will be structured and systematic and will address all the necessary aspects of the organization to permit the safe and responsible use of any UAS within the organization or enterprise. Missing, ignoring, or misjudging any aspect may ground the entire program before its first flight.

Outsource

Alternatively, outsourcing has its own list of benefits and potential drawbacks. Some obvious benefits include speed, deferred / shared risk, mission / project specific UAS, near real-time access to content, and the list can go on and on. All of this comes with a premium, financially speaking. Also, it may have latency or other delays since an outside firm will provide the service. Contracts will need to be in place and service level agreements (SLAs) established, and liability / risk matters clearly defined. When services are outsourced, a trade-off between flexibility and control often comes with the territory. In this case, flexibility may come in the form of speed – rapid acquisition of content or imagery, a specialized type of platform or sensor for a specific need, current technologies, and many others.

Outsourcing allows an organization to rapidly acquire the benefits of a technology or service without incurring the upfront capital expenditures and time associated with establishing the technology or service internally. This allows the organization to quickly assess and demonstrate the real benefit to the organization while providing an avenue to quickly change course should the benefits fail to materialize or should new solutions enter the market. Once the benefits become economical, the organization can develop a plan or strategy to bring these services or technologies in-house in a structured and systematic manner. Conversely, if the benefits are sporadic, customer specific, or too complex to adopt / integrate it may be advisable to continue with the outsourcing approach for extended periods of time.

Many companies to reduce overhead, maintain competitive advantage, and remain focused on core lines of business have used long-term outsourcing of business critical activities successfully. If not managed properly the best outsourcing strategy may introduce "drag" or "friction" into the business. Good stewardship and governance are cornerstones of all good outsourcing strategies.

Governance

Whether an organization decides to insource or outsource, all aspects of a UAS program will need to have a governance framework to manage and control risks, liabilities, and operations. When using a UAS for any commercial enterprise, many internal and external policy, regulatory, operations, and contractual matters require strict conformity standards and controls. All data collected using any UAS platform will need to be handled in accordance with clear controls, policies, and operating procedures for the collection, processing, distribution, consumption, access, and storage/retrieval of data.

For example, ACME Antenna decides to use a UAS to inspect remote towers that provide cellular, microwave, radio and other signal types for multiple customers. ACME anticipates an increase in productivity due to the speed at which a UAS can observe and document towers with fewer resources and no human exposure to high-risk environments (people climbing towers). Through the use of this technology,

ACME expects the maintenance team to be deployed more effectively, which will have the added benefit of improving the quality and reliability of their service. Since tower workers will be spending less time climbing the towers for inspections, ACME also expects to reduce their insurance premiums, improve worker morale, and reduce workmen compensation claims.

Initially, ACME is going to test this theory by outsourcing a select number of towers in multiple geographic locations that represent a cross section of the towers in the ACME portfolio. They will collect UAS data for a period of one year with periodic executive reviews of the program. The company selected to perform the work, Aerial Strategies, is experienced and has all the necessary legal and regulatory compliance paperwork in place to conduct the work. They are the perfect company to perform the work.

Aerial Strategies sets off to conduct aerial inspections in accordance with the ACME contract. They fly using the latest and greatest technologies and capture photos, IR, thermal, and multispectral data of each tower. The photos are high resolution and are sufficient detail for review and assessment by an authorized ACME engineer. Thermal, IR, and multispectral data will potentially show stress, fatigue, or other maintenance or production issues that may jeopardize service. The collected data is processed by Aerial Strategies in accordance with the guidelines and standards provided by ACME. This will permit the easy assimilation of the data into the ACME infrastructure. Each tower has an average of 40 high-res photographs, each approximately 10MB each (400MB for pictures only), and 10 IR, thermal, multispectral (approx. 100MB in total).

Each tower inspection is an average of 500MB of data and they plan to conduct 1000 inspections during this testing period. This will translate to approximately 500GB of data for the tower inspections. All this data will need to reside on the ACME network and will be subject to any and all corporate data governance policies, controls, and standards. In order to process and host this data, ACME may need to allocate additional equipment to their Information Technology (IT) infrastructure to process, store, and access this data during the testing period. During the testing period, it may be warranted to have dedicated resources for this effort in order to segregate this data from other business critical data. Procedures for ingesting, processing, storing, accessing and enriching this data will need to be established, monitored, and controlled during the test period.

The test data will need to be accessible by hundreds of internal and external people distributed across multiple geographic locations on two continents. The enriched data will provide even better business intelligence information, which is anticipated to reduce their annual maintenance budget by 20% due to unnecessary maintenance trips and reduce insurance claims by 30% due to a 50% reduction of high-risk manned inspections. The expectations, although high, are promising and attainable. This could revolutionize their industry and propel ACME Antenna into a leadership position within their industry.

The testing period begins and ACME soon discovers through the thermal and multi-spectrum data that many of their tower components are fatigued and are likely to fail well before their planned replacement. This finding is soon confirmed by analyzing historical data. ACME reviewed trouble tickets from the past five years and discovered that incidents of service disruptions and outages, as well as maintenance repairs, had been steadily on the rise. Furthermore, the specific components that were in decline were caught well in advance of failing. When a component fails all evidence of its condition is lost due to the destruction that occurs during failure. With this data, ACME is now able to work with component manufacturers to address quality issues of the components. The component provider quickly identifies and remedies the component deficiency, and service disruptions and outages are immediately improved.

This one change alone had the added benefit of reducing service calls to the call center, which allowed the call center to improve their ticket resolution numbers. This unanticipated savings was not only something that could be dollarized, ACME was soon receiving positive feedback from customers regarding their service. Soon after, ACME began seeing the number of new customers increase three months after these changes were implemented. Six months into the testing period the inspection program had already paid for itself and new avenues for using drones were being conceived.

RISKS & LIABILITIES

We can't talk about all these fantastic benefits and opportunities without discussing and addressing the risks associated with new flying technologies. Is there risk? Absolutely! Can it be managed? Yes! With the introduction of any new technology – be it the car, telephone, computers, airplanes, horse and buggy – no matter how far back you go in history, you will find that the initial reaction to all new revolutionary technologies was that of pessimism. It was common for society, the media, and the governing bodies to first respond with fear, uncertainty and doubt (the proverbial FUD Factor). People are inherently slow to accept change, and the most basic reason is based on fear or risk.

The Drone Age is no different. Anytime you take something and make it capable of flight the risks increase. I don't think anyone would argue that. Variables increase by magnitudes once something is no longer tethered to the ground, or even when it is, the risks increase. When the airplane was introduced, and even when it was widely acceptable, it was extremely risky. Early commercial airplanes had little or no navigation aids. It was often based on the experience and judgment of the pilot and good ol' dead reckoning (flying in a known direction for a specific period of time) to reach a destination. Even with the best pilots, the planes themselves were unreliable and had failure rates that would not be acceptable today. Only the most adventurous travelers of today would venture to embark on a trans-continental or trans-Atlantic flight on an old plane.

With the advent of something new, we assume the risks are high since there is little or no actual historical data to tell us it is safe. I often speak about risk is terms of safety, security, and privacy and call them the Three Pillars Risk. Safety is a big deal for everyone, and for good reason, and we hastily adopt a "never" or "No way" stance and apply it to everything associated with the risk, in this case, a commercial UAS (drone). When we begin to break down risk so we can better understand it, then we begin to loosen our "never" or "No way" position. This is beginning to happen now in the U.S.

When we think about risks to people, we say, okay this has to be of paramount concern, and everyone agrees. When we think about the risk to property, we also start formulating what may happen if a drone hits a concrete pillar (we lose a drone and MAYBE scratch the concrete,) or hit a car, or something more or less significant. Once we adopt a risk-based approach to the use of a drone then we can start establishing policy and rules for their use based on the risk they pose to people and property. Can we fly them in the middle of the desert with little risk? Yes! Can we fly one over a stadium full of people with little risk? Not currently. So the spectrum is rather broad and we need a regulatory policy and operating standards that allow safe and responsible use to occur.

Also, we need a prescription and remedy (cause / effect) framework and process that permits the systematic handling exceptions and the inevitable consequences associated with things going wrong. Fortunately, the U.S. and all other developed countries already have robust regulatory and statutory processes for handling all manner of risks and accidents. The UAS industry does not require mountains of laws and regulatory friction. We do need to account for them for sure. I am a strong advocate for the safe and responsible use of any/all UAS platforms in U.S. airspace. Though I have seen cases where the regulatory posturing has gone so far off the rails that a paper airplane would be considered a UAS (drone)and, therefore, would need to comply with FAA regulations and guidelines.

Risk will persist on many levels: operating risk, organizational, personal, political, headline, policy, human, and many others. All of them are due proper consideration and evaluation.

Fortunately, the world is full of expert risk assessors and managers; insurance companies do this everyday. Several are jumping into the commercial UAS industry and are establishing frameworks that are being used in the UAS industry today. As more data is available these risk tools will become more comprehensive and accurate and thus more meaningful to the entire industry. Managing and avoiding risk is foundational to the safe operation of any aerial business, whether it is a manned or unmanned aircraft. With risk comes liability. The commercial UAS industry is still in its infancy, and for many business operations liability is considered a barrier of entry for many well-established companies. Many companies are reluctant to put their entire business at risk when the liability may be higher than the benefit of using any UAS platform.

The future is bright and new capabilities, applications and benefits are realized everyday. The proverbial genie is out of the bottle and it will not and cannot be put back in.

THE FUTURE OF THE UAS REVOLUTION

PART 4

WHAT'S NEXT, WHERE DO WE GO FROM HERE: FUTURE STATE

When we think about unmanned vehicles (UV) and unmanned aerial System (UAS) specifically we should keep an open mind to all the possibilities that are yet to be discovered. With respect to unmanned vehicles, we have already explored our oceans, distant planets and even have one (Voyager I) that left our solar system and is exploring interstellar space. It shouldn't be too far of a stretch to imagine the cohabitation of manned and unmanned aircraft safely working in our skies.

We can extrapolate further and look at Moore's Law, which has held true since it was published in 1965. The original premise was that the amount of transistors that could be placed into an integrated circuit would double every two years. Technology has moved from transistors to a purely digital space, and Moore's Law remains true with the digital age. How would Moore's Law apply to The Drone Age? I think it is safe to say that the number of UAS platforms in our skies is going to increase dramatically over the next 5-10 years, probably by several magnitudes. Attitudes will change, adoption will increase and the benefits will be many and varied.

The fantastic realities of what this means in terms of applications and how drones will be used is fascinating to consider. We know miniaturization of digital components will continue, processing power will increase in speed and decrease in size, sensors will continue to get smaller and more powerful, battery and solar technology will continue to advance in amazing ways. If we start thinking about how the convergence of all this applies to drones then our imaginations are free to explore possibilities.

We already have the idea of fully autonomous flight and swarms and perhaps even fleets. Swarms coordinate their activities around a common purpose or goal (collect IR data for crop management). A fleet would be comprised of several swarms, each swarm having a designated purpose or mission. Let's consider a fleet of delivery swarms, perhaps one or many swarms are delivering individual packages to a region, the fleet is centrally managed and coordinated, while the swarms operate more autonomous and in compliance with their stated purpose: to deliver a package. The fleet, however, may have other tasks along its route that may involve coordination or collaboration with other fleets or swarms – the fleet also is inspecting major infrastructure along its route for a new customer. Perhaps the fleet can be tasked in real-time and it manages the swarms or excess capacity within the fleet. Once a swarm or individual drone completes its task it could return to the fleet for additional or new tasking. The possibilities are amazing to ponder.

Now when combining a UAS with other amazing technology such as quantum computing, nanotechnology, and Artificial Intelligence, the possibilities get futuristic and perhaps a little scary. If I were writing a sci-fi novel, I could conjure up some pretty scary possibilities. In the realm of reality and possibility, we can look at many other real opportunities for advanced aerial autonomous platforms to operate in an integrated airspace in concert with manned aircraft.

Sense and avoid technology for UAS is already happening, though still primitive. With Moore's Law, we can predict that commercial viability is only three to five years away. Couple that technology with existing aviation transponders and advanced capabilities and the realities look quite compelling. The FAA is working on the next generation (Next-GEN) air traffic control system which uses GPS navigation data and permits aircraft to broadcast their direction, speed, and other pertinent data to other aircraft in the area without coordination with air traffic control (ATC) on the ground. This is a great system and will improve the safety of our airways a great deal.

I had the pleasure of flying a Cessna 206T from the west coast to the east coast, and it was outfitted with this capability. When it received the signal (ADS-B) from another aircraft the information was displayed in my flight management system. It was amazing and incredible to see and experience. This capability can be and should be made available to the UAS industry for all large and small UAS platforms. This would remove a significant amount, if not all, manageable risk out of the system. It can't fix malicious actions, but it can reduce or eliminate potential accidents. That is a great idea and the technology is already available for that solution.

Sense and avoid, integrated and coordinated in the national airspace system (NAS), autonomous flights (fleets & swarms), then what's next? Let's play a little. Let's throw some AI, quantum computing, and nanotechnology in the mix. Stay with me.... I know what you're thinking.... sci-fi. Nope. Let's use our fleet and combine it with AI and quantum computing. Our fleet has an endurance of two to three hours and is comprised of several swarms, each with a specific capability. One has imaging (visual and non-visual spectrums – IR, thermal, hyper spectrum, LiDar, and several chemical and radio/electromagnetic sensors); one swarm has robotic arms with various capabilities (dig, scoop, and shoot with a laser). The last swarm has nanotechnology capable of creating 60% of all known chemical compounds. Stay with me – don't get scared....

With these capabilities, the fleet could fly over any terrain and use its capabilities to monitor, inspect, and perhaps treat or repair any pre-programmed conditions. If we keep the flight speed to 50 miles per hour and an altitude of less than 1000 feet we will have a rather large coverage area. Let's say we unleash this fleet over an oil spill with the mission to contain and disperse it. With the AI, quantum computing and nanotechnology, the fleet could autonomously analyze the scenario and determine the area and density of contamination immediately. With its robotics, the drones could test and document the exact chemical makeup and devise an immediate treatment program, then create the remedy using nano-technology to counteract the oil spill. If the fleet did not have enough of anything, it would summon additional specialized swarms until the mission was complete. I know that seems pretty scary to some, I am sure. It may not be as far into the future as you think!!

Now imagine another fleet doing the same kind of thing, but it is flying in the stratosphere and is using solar power and has an endurance of six months or even unlimited. Its tasking is to analyze and remove ozone from the atmosphere. Using all the same quantum and nano-technology, it would fly continuously and be replaced or augmented with new units as necessary to maintain a constant vigilance in our atmosphere, working to undo all negative impacts associated with elevated levels of CO_2 in our ecosystem.

I can imagine scenarios where UAS small and large will be so commonplace that the fears and concerns we have today will be long forgotten and overshadowed by the benefits they offer us. Society was skeptical of all major advances in the past, and if the past is any indicator of the future, then I feel confident saying that this industry will be vibrant and long lasting.

Today we use unmanned space aircraft to shuttle cargo to/from the International Space Station (ISS). That too was shunned and doubted not long ago. The age of the self-driving car was that of science fiction and reserved for futuristic movies. Well, today they are becoming more than fantasy; some countries are adopting driverless vehicles for mass transit applications. If we look further we will learn that many unmanned vehicles are more pervasive than we think. Under our oceans unmanned vehicles have replaced many, if not most or all, high-risk inspection activities once performed by humans. These unmanned vehicles can remain in the hazardous environment for hours or days, whereas a human may only be able to stay for minutes.

Law enforcement has been using unmanned vehicles or "robots" for years to aid them in high hazard environments where explosives, biological agents, or any other dangerous situations are present. The skies are no different. It may take longer to ensure the safe operations of such vehicles, but the reality is that it is just a matter of time. In fact, most large commercial aircraft are already capable of flying and landing and coming to a stop without intervention from a pilot

The future is full of fantastic applications and benefits and we are truly only limited by our imaginations. I, for one, am excited about the wondrous ways these mighty powerhouses of possibilities will be employed to the benefit of humanity and our precious planet. Like so many, if not all, technological advances before this, we can say with confidence we ain't seen nothing yet! The sky's the limit!!

CPSIA information can be obtained at www.ICGtesting.com
Printed in the USA
LVIW01n1419020616
490957LV00016B/247